Series on

Reproduction and Development in Aquatic Invertebrates

Volume 6

Reproduction and Development in Minor Phyla

Series on

Reproduction and Development in Aquatic Invertebrates

Volume 6

Reproduction and Development in Minor Phyla

T. J. Pandian

Valli Nivas, 9 Old Natham Road
Madurai-625014, TN, India

CRC Press
Taylor & Francis Group
Boca Raton London New York

CRC Press is an imprint of the
Taylor & Francis Group, an **Informa** business

A SCIENCE PUBLISHERS BOOK

Cover page: Representative examples of minor phyletic species. For more details, see Figure 1.1.

First edition published 2021
by CRC Press
6000 Broken Sound Parkway NW, Suite 300, Boca Raton, FL 33487-2742

and by CRC Press
4 Park Square, Milton Park, Abingdon, Oxon OX14 4RN

Library of Congress Cataloging-in-Publication Data

Names: Pandian, T. J., author.
Title: Reproduction and development in minor phyla / T.J. Pandian.
Description: First edition. | Boca Raton : CRC Press, 2021. | Series:
  Series on reproduction and development in aquatic invertebrates ; volume
  6 | Includes bibliographical references and index.
Identi iers: LCCN 2020035992 | ISBN 9780367523350 (hardcover)
Subjects: LCSH: Invertebrates--Reproduction. | Invertebrates--Development.
Classi ication: LCC QL364.15 .P36 2021 | DDC 592--dc23
LC record available at https://lccn.loc.gov/2020035992

ISBN: 978-0-367-52335-0 (hbk)
ISBN: 978-0-367-52336-7 (pbk)

Typeset in Palatino
by Radiant Productions

Preface to the Series

Invertebrates surpass vertebrates not only in species number but also in diversity of sexuality, modes of reproduction and development. Yet, we know much less about them than we know of vertebrates. During the 1950s, the multi-volume series by L.E. Hyman compiled some information on reproduction and development of aquatic invertebrates. Through a few volumes published during the 1960s, A.C. Giese and J.S. Pearse provided a more concrete shape to the subject of Aquatic Invertebrate Reproduction. During the 1990s, K.G. Adiyodi and R.G. Adiyodi in their multi-volume series on Reproductive Biology of Invertebrates elevated the subject to a visible and recognizable status.

Reproduction is central to all biological events. The life cycle of most aquatic invertebrates involves one or more larval stage(s). Hence, an account on reproduction without considering development would remain incomplete. With the passage of time, publications are pouring through a large number of newly established journals on invertebrate reproduction and development. The time is ripe to update the subject. This treatise series proposes to (i) update and comprehensively elucidate the subject in the context of cytogenetics and molecular biology, (ii) view modes of reproduction in relation to Embryonic Stem Cells (ESCs) and Primordial Germ Cells (PGCs) and (iii) consider cysts and vectors as biological resources.

Hence, the first chapter on Reproduction and Development of Crustacea opens with a survey of sexuality and modes of reproduction in aquatic invertebrates and bridges the gaps between zoological and stem cell research. With the capacity for no or slow motility, the aquatic invertebrates have opted for hermaphroditism or parthenogenesis/polyembryony. In many of them, asexual reproduction is interspersed within sexual reproductive cycle. Acoelomates and eucoelomates have retained ESCs and also reproduce asexually. However, pseudocoelomates and haemocoelomates seem not to have retained ESCs and are unable to reproduce asexually. This series provides a possible explanation for the exceptional pseudocoelomates and haemocoelomates that reproduce asexually. For posterity, this series intends to bring out six volumes.

August, 2015   T. J. Pandian
Madurai-625 014

Preface

The 26 recognized minor phyla comprise aberrant clades that are usually not considered in the mainstream of evolution, as most of them terminate as blind offshoots. There are a large number of hypotheses and theories describing their phylogenetic relation and affinity. Discussions and debates on them have been ongoing endlessly. A vast majority of publications, reviews, chapters, books and book series elaborate the inter-relations between minor phyla and the intra-relation to major phyla. As there is more than adequate information on this theme, this book shall not describe or discuss their phylogenetic relations. For the first time, this book is entirely devoted to focus on other aspects, which have not so far been sufficiently recognized. It represents a comprehensive synthesis of relevant information on reproduction and development in minor phyla.

The book is arranged in 26 chapters, which are grouped into 6 parts, each of which comprises phyla classified on the basis of coelom. The first chapter highlights some interesting features of the minor phyletics. Chapters 2 to 4 under Aorganomorpha bring to light the presence of limited number of cells and cell types (4–7) in free-living Placozoa, and parasitic Mesozoa and Myxozoa. In the last two, gametogenesis and sporulation occur within a single cell, providing a unique opportunity for cytologists to study the rare process of endogony. The next four chapters summarize the limited information available for Acoelomorpha consisting of Loricifera, Cycliophora, Nemertea and Gnathostomulida.

The third major clade Pseudocoelomata include the speciose Rotifera and Nematoda, and the relatively less speciose Gastrotricha and Kinorhyncha as well as the parasitic Nematomorpha and Acanthocephala. Unlike trematodes and cestodes, nematodes are parasitic on plants and animals, and inflict heavy losses to man and his food basket. The nematomorphs are parasitic as juveniles but free-living as adults.

Another major clade Hemocoelomata also includes six minor phyla; of them Priapulida, Sipuncula and Echiura are exclusively marine inhabitants. Tardigrada and Onychophora are limnic, limno-terrestrial or terrestrial inhabitants; along with Pentastomida, the last two are characterized by segmentation and molting either periodically or occasionally. In priapulids and possibly others also, eutelism limits fecundity to < 1,000 eggs.

Environmental switching of male sex differentiation in *Bonellia viridis* is elaborated.

The fifth major clade Lophophorata consists of Entoprocta, Bryozoa, Phoronida and Brachiopoda. Of them, the first too are colonials and clonals. The lophophorates are exclusively marine inhabitants except for the limnic Urnatellida among entoprocts and Plumatellida among bryozoans. Whereas the eucoelomatic Ascidiacea and Chaetognatha are hermaphrodites, the hemichordates are gonochores. Surprisingly, the colonial bryozoans (5,700 species) and ascidians (3,000 species) are more speciose than the solitary chaetognaths (150 species).

The minor phyla are not as speciose (1,795 species/phylum) as the major phyla (157,066 species/phylum). Gonochorism obligately requires motility to search for a suitable mate. The combination of low motility and gonochorism from Placozoa to hemocoelomatic minor phyla has limited the minor phyletic species diversity. The accumulation of deleterious genes causes inbreeding depression among progenies arising from parthenogenesis or selfing hermaphrodites and/or clonal multiplication. In eutelics, mitotic division is ceased in somatic cells after hatching. For the first time, the eutelism and its prevalence is brought to light in all the six pseudocoelomatic phyla. In the eutelics, the inability of limited number of somatic cells to support vitellogenesis and the like in gamete maturation has limited fecundity to 30–300 eggs in free-living pseudocoelomates, priapulids and possibly other hemocoelomates. In them, sperm production is more limited than that of egg production; consequently, a large number of eggs remain unfertilized. By limiting the fertilized egg and progeny number, the scope for species diversity is also limited. Nevertheless, a high proportion of non-eutelic gametic cells (35%) in nematodes and possibly other parasitic pseudocoelomate phyla facilitate more fecundity than in rotifers, in which the proportion of the non-eutelic gametic cell is 15%. Briefly, the reasons for the limited species diversity in minor phyletics are traced to (i) eutelism (65.7% of all minor phyletic species), (ii) clonal multiplication (21.6% of all minor phyletic species), (iii) parthenogenesis (6.4% of minor phyleitc species) and (iv) ~ 1% selfing hermaphrodites. With adoption of an array of strategies, the number of selfing minor phyletic species is restricted to < 100 species (e.g. dicyemid Mesozoa). All others have successfully eliminated the scope for inbreeding depression. Yet, the need to manifest and maintain dual sexuality, fecundity in 19% of hermaphrodites may be reduced to 50% of that in gonochores. This deficiency seems to have been compensated by supplementation of clonal multiplication in colonial bryozoans and ascidians. Yet, how can the non-selfing, hermaphroditic bryozoans (5,700 species) and ascidians (3,000 species) with clonal multiplication and consequent potential inbreeding depression be more speciose than their gonochoric counterparts? For the first time, the obscure and unintentional reports on the fusion between fragments and/or colonials—an event equivalent to gametic fusion at fertilization—are brought to light as the source of new gene combination.

Besides, their colonies degenerate and regenerate more or less regularly. Only fittest degenerated colonies may be rejuvenated and regenerated. Thus, species richness in these colonials is traced to fragmental/colonial fusion and occurrence of regeneration of the fittest colonies.

This book is a comprehensive synthesis of publications by 560 authors covering > 800 minor phyletic species, selected from widely scattered 334 journals, books and other sources. The holistic and incisive analyses have led to several new findings related to reproduction and development in minor phyletics and to project their uniqueness. Hopefully, this book serves as a launching pad to further advance our knowledge on reproduction and development in the minor phyletics.

April, 2020   T. J. Pandian
Madurai-625 014

Acknowledgements

It is with pleasure that I thank Dr. E. Vivekanandan and P. Murugesan for critically reviewing parts of the manuscript of this book and for offering valuable suggestions. In fact, I must confess that I am only a visitor to the theme of this book. However, my editorial services on energetics of some Minor Phyla (Pandian, 1987, Animal *Energetics*, Academic Press) has emboldened me to author this book. Thankfully, The American College, Madurai lent me classical text books. The manuscript of this book was prepared by Mr. T.S. Surya, M.Sc. and I wish to thank him profusely for his competence, patience and co-operation.

I wish to thank many authors/publishers, whose published figures have been simplified/modified/compiled/redrawn for an easier understanding. To reproduce original figures from published domain, I gratefully appreciate the permission issued by S. Segers. I welcome and gratefully appreciate the open access policy of PLoS ONE, Development, Growth and Differentiation and Parasites & Vectors. For advancing our knowledge in this area by their rich contributions, I thank all my fellow scientists, whose publications are cited in this book.

April, 2020   **T. J. Pandian**
Madurai-625 014

Contents

Part C: Pseudocoelomata

Part D: Schizocoelomata: Hemocoelomata

Part E: Schizocoelomata: Lophophorata

Part F: Eucoelomates

1

General Introduction

Introduction

The minor phyla comprise aberrant clades that are not usually considered in the mainstream of evolution. In the phylogenetic tree, most of them terminate as blind offshoots, albeit a few are regarded as a link between two or more major phyla; for example, the velvet worm Onychophora provides a vital link for arthropod evolution. Hyman (1951a, b, 1959) has brought out excellent accounts for 16 minor phyla among the then known 22 aberrant clades. As new taxa are being continuously erected, four minor phyla (Placozoa, Myxozoa, Loricifera, Cycliophora) are now added. Pogonophora is excluded and is now included in polychaetes (see Pandian, 2019). Presently, the aberrant clades are classified into 26 minor phyla. Figure 1.1 shows perhaps for the first time the pictures assembled for representative species from 26 minor phyla. There are a large number of hypotheses and theories describing the phylogenetic relations and affinity between two or more phyla. Discussions, and debates are ongoing on them. A vast majority of publications, reviews, chapters, books and book series elaborate the relationships and affinity between these phyla. As there is more than adequate information on this theme, this account shall not describe or discuss the phylogenetic relations. But it is rather devoted to focus on other aspects, which are not thus far adequately recognized.

Taxonomic Distribution

The minor phyla are not as speciose (1,795 species/phylum, Table 1.1), as the major phyla are (157,066 species/phylum). In six of them, the species number is < 100 (Placozoa, Loricifera, Cycliophora, Gnathostomulida, Priapulida, Phoronida), a few hundred in a dozen of them (Mesozoa, Gastrotricha, Kinorhyncha, Nematomorpha, Sipuncula, Echiura, Onychophora, Pentostomida, Entoprocta, Brachiopoda, Chaetognatha, Hemichordata)

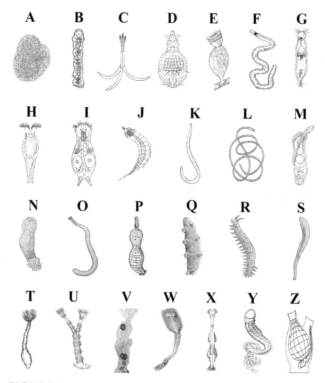

FIGURE 1.1

A. Placozoa: *Trichoplax adhaerens* (Eitel et al., 2011), B. Mesozoa: *Dicyema japonicum*. C. Actinospore of Myxozoa. D. Loricifera: *Rugiloricus*, E. Cycliophora: *Symbion pandora*, F. Nemertea: *Ototyphonemertes* sp (after Botosaneanu, 1986), G. Gnathostomulida: *Gnathostomula jenneri* (drawn from Biocyclopedia), H. a bdelloid rotifer, I. Gastrotricha: *Chaetonotus* sp (from Biology Discussion), J. Kinorhyncha: *Echinoderes* sp (free hand drawing from Bumblebee.org), K. Nematoda: *Ascaris lumbricoides*, L. Nematomorpha: *Gordius* (from Biology Discussion), M. an acanthocephalan (from NC State University), N. a priapulid (after Kingsley, 1884), O. Sipuncula: *Sipunculus nudus* (drawn from Maiorova and Adrianov, 2017), P. Echiura: *Echiurus* (drawn from Biocyclopedia), Q. a tardigrad (from New York Times), R. Onychophora: *Peripatus* (drawn from BioScience Media), S. Pentastomida: *Linguatula serrata* (drawn from Saari et al., 2019), T. a phoronidan (drawn from Spektrum.de), U. Entoprocta: *Urnatella* (from Biology Discussion), V. a bryozoan (drawn from Academic Web Site), W. Brachiopoda: *Lingua* (drawn from Udo Savalli, Arizona University), X. Chaetognatha: *Sagitta elegans* (from New World Encyclopedia), Y. *Balanoglossus* Z. solitary ascidian (all others are free hand drawing from Hyman, 1959).

TABLE 1.1

Number of phyla and approximate species number in major (from different sources) and minor phyla (from this book)

Major Phylum	Species (no.)	Minor Phylum	Species (no.)
Aorganomorpha			
Porifera	8553	Placozoa	3
Cnidaria	10856	Mesozoa	150
Acnidaria	166	Myxozoa	2200
Acoelomorpha			
Platyhelminthes	21810	Loricifera	34
Turbellaria	5500	Cycliophora	2
Non-Turbellaria	16310	Nemertea	1300
		Gnathostomulida	100
Psuedocoelomata			
Annelida	16911	Rotifera	2031
Echinodermata	7000	Gastrotricha	813
		Kinorhyncha	200
		Nematoda	27000
		Nematomorpha	360
		Acanthocephala	1100
Hemocoelomata			
Arthropoda	1166984	Priapulida	19
Crustacea	54384	Sipuncula	160
Non-crustacea	1112600	Echiura	230
Mollusca	118451	Tardigrada	1047
		Onychophora	200
		Pentastomida	144
Lophophorata			
		Entoprocta	200
		Phoronida	23
		Brachiopoda	391
		Bryozoa	5700
Eucoelomata			
Vertebrata	62869	Chaetognatha	150
		Hemichordata	130
		Urochordata	3000
Major phyla total	1413600	Minor phyla total	46687
Grand total		1460287	
157,066 species/major phylum; 1,795 species/minor phylum			

and ranges from 1,000 to 3,000 in another six of them (Myxozoa, Nemertea, Rotifera, Acanthocephala, Tardigrada, Urochordata). Only Nematoda (27,000 species) and Bryozoa (5,700 species) are speciose. Further, a few phyla are represented by two-three species (e.g. Placozoa, Cycliophora) and two families (e.g. Priapulida, Onychophora). Some of these values may be compared with the least speciose Echinodermata (7,000 species, Pandian, 2018) and > million speciose Arthropoda (Chapman, 2009). Presently, the minor phyla comprise 46,687 species. They are grouped into (i) Aorganomorpha (a new term coined to indicate the lack of organ, in line with acoelomorpha with no coelom), (ii) Acoelomorpha, (iii) Pseudocolomata, (iv) Hemocoelomata, (v) Lophophorate schizocoelomata and (vi) Eucoelomata. The number of phyla within each of these groups and the number of species in each phylum are shown in Fig. 1.2. Among them, pseudocoelomates with six phyla, each with the highest mean species number of 5,251 per minor phylum, have generated more often aberrant clades. Surprisingly, they are all eutelic, i.e. after hatching, mitosis is ceased in their somatic cells, a feature which has not been adequately recognized. Notably, major phyla have no representative from Pseudocoelomata.

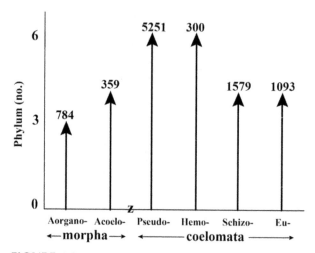

FIGURE 1.2

Number of phyla and species within the major clads of minor phyla. Data are drawn from Table 1.1.

Phylogenetic Cladograms

Enormous efforts have been made to construct phylogenetic cladograms to track the history of animal evolution. Based on structural and embryological features, zoologists have traced the animal evolution from simple to complex animals through gradual steps and bridging links, although they are very much aware that parasitism, for example, reverses the course from a complex to simple form, as in parasitic rhizocephalic (e.g. *Sacculina carcini*) crustaceans (Pandian, 2016). For classification, the conventional cladogram considers the absence of coelom, and the presence of pseudocoelom, hemocoelom, lophophorate schizocoelom and (enteric) eucoelom (Fig. 1.3A). On the other hand, molecular biologists use ribosomal *RNA* genes and others (Littlewood et al., 2015) to construct a molecular cladogram (Fig. 1.3B). A major proportion of these classifications are common to both morphological and molecular cladograms. Both of them classify the diploblastic radial and triploblastic bilateral symmetry in animals. They also recognize the distinct difference between Protostomia and Deuterostomia. In the former, the blastopore differentiates into a mouth but into an anus in the latter. The difference appears between these two cladograms, once the morphological cladogram classifies coelomates into Dueterostomia, Lophophorata and Protostomia,

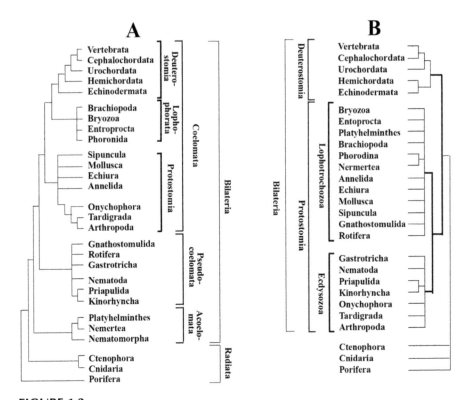

FIGURE 1.3

Metazoan phylogenies. A. The traditional phylogeny based on morphology and embryology. B. The new molecule (rRNA)-based phylogeny (compiled, modified and redrawn from Hyman, 1940, Adoutte et al., 2000).

whereas the molecular cladogram classifies it into Lophotrochozoa and Ecdysozoa. The Ecdysozoa are characterized by the presence of cuticle, which is periodically molted during the life cycle. Within it, Tardigrada, Onychophora and Arthropoda are grouped into Panarthropoda. The other Cycloneuralia includes Nematoida (with Nematoda, Nematomorpha) and Scalidophora (with Priapulida, Kinorhyncha, Loricifera) (Blaxter and Koutsovoulos, 2015, Hejnol, 2015b).

Both the cladograms are neither complete nor totally reliable. For example, Chaetognatha is not placed in the molecular cladogram. Despite the ever-increasing number of molecular studies, chaetognaths are an enigmatic clade in triploblastic metazoan phylogeny (Perez et al., 2014). Regarding morpho-embryology cladogram, "the cleavage patterns of the primitive invertebrates namely Placozoa (Grell, 1972), Orthonectida (Caullery and Lavallee, 1908a, b), Porifera (Fell, 1974) and Cnidaria (Uchida and Yamada, 1983) are so diverse that a similarity in cleavage pattern *per se* may not necessarily reflect phylogenetic affinity between these clades" (Furuya et al., 1996). A few examples may indicate the inadequacy of information on molecular tags and contradictory conclusions, when different molecular tags are used. Jensen and Bullard (2010) collected 25 cestode species from elasmobranchs and 27 larval cestode species from the Gulf of Mexico. Using a molecular tag, they were able to place only some of them and that too only up to the genera. With limitation of resolving power of a molecular tag (Alvarez-Presas and Riutort, 2014), a combination of many tags has to be assembled to construct a molecular cladogram for turbellarians (Pandian, 2020). Studies on the orthonectid Mesozoa revealed that they are degenerated annelidan worms (Schiffer et al., 2018), when 18S rDNA tag was used but their affinity is closer to ciliate protozoans, when 5S rRNA tag was used (Furuya et al., 1996). To the credit of molecular markers, however, the following may be noted: molecular tags have been successfully used to describe the complicated parasitic life cycle as well as to trace phylogenetic distribution of trematodes that use mollusks and fish as first and second intermediate hosts. For the first time, Huston et al. (2016) successfully employed ITS2 rDNA sequences generated for the trematode larvae of *Gorgocephalus yaaji* from the infected snail *Echinolittorina austrotrochoides* to describe the complete but complicated life cycle of a trematode. Using 18S rDNA sequence, Cribb et al. (2001) inferred the phylogeny and distribution of ~ 51 digenean superfamilies that employ gastropods (30 species), bivalves (20 species) and scaphopod (1 species) as intermediate hosts as well as 56 orders of fish that serve as definitive hosts.

Amazing Features

In some of these minor phyletic species, a few amazing features must be mentioned. (i) *Habitats*: Some gastrotrichs are found abundant for extended periods (months) in anoxic habitats, in comparison to *Artemia* cyst, which can withstand anoxia for 4 years (y) (see Pandian, 2016). The gastrotrichs seem to possess a sulfide detoxification mechanism similar to that demonstrated for freshwater nematodes (Strayer et al., 2010). Abysmal depth has not eliminated the existence of cuticularized loriciferans from 480 m depth (Krisetensen, 1983), priapulids 2,500 m depth (Adrianov and Malakov, 1996), kinorhynchs 5,200 m depth (Neuhaus and Higgins, 2002) and soft-bodied echiurans 9,000 m depth (Dawydoff, 1959). Earlier the shelled buccinids and prosobranchs alone were collected from 8,000–9,050 m depth in the Kuril Trench (see Pandian, 2017). (ii) *Occurrence*: Being microscopic in size, some gastrotrichs occur at density of 2.6 million/m^2 in the Polish Lakes (Strayer et al., 2010), nematodes 38.2 million/l (Glime, 2017c) and rotifers 1 million/l in natural waters and up to 10^7 individual/l in culture system (Wallace and Snell, 2010). Incidentally, the reported highest density for copepod cyst is 3.2 million/m^2 for *Acartia clausi* (see Pandian, 2016). (iii) *Temperature*: Amazingly, the highest temperature, at which a few nematodes collected from hot springs, ranges from 52.0°C for *Rhabdolaimus brachyuris* in Italy to 57.6°C for *Aphelenchus* sp in Chile and to 61.3°C for *Aphelenchoides* sp in New Zealand (Poinar, 2010). On the other hand, the tardigradan boar worm *Bertolanius nebulosus* can supercool and yet survive at –7°C; *Ramazzottius oberhaeuseri* remains indestructible down to –80°C (see Glime, 2017b). Some of these values may be compared with the ability of fish like African tilapias inhabiting hot soda lakes at 44°C and *Trematomus* found under the Antarctic ice sheets at –2°C (see Pandian, 2010). (iv) "*Water* is life's matter and matrix, mother and medium. There can be no life without water" (Glime, 2017b). Astonishingly, some tardigrades are credited with a century-long survival in a cryptic almost totally dry state. From a herbarium moss specimen holding a tardigrade, the bear worm began cellular activity after 120 years (see Glime, 2017b). (v) *Resting egg* and *cyst*: cladocerans and copepods produce resting eggs, which survive for several decades and centuries in a non-desiccated form (Clark et al., 2012); for example, the longest viable duration ranges from 125 years for the cladoceran hydrated ephippia to > 332 years for the copepod resting egg (e.g. *Diaptomus sanguineus*, see Pandian, 2016). Under anhydrobiotic state, *R. oberhaeuseri* eggs may, however, remain viable only for 9 years (Glime, 2017b). (vi) *Sexuality*: In bdelloid rotifers, the presence of the male has not thus far been reported. Nevertheless, they have persisted and diversified into 461 species during the last 60 million years of their existence, challenging evolutionary biologists for an explanation. (vii) *Life history*: Idiosyncratic piliciloricidan loriciferans are acoelomates as larvae but pseudocoelomates as adults (Kristensen, 1995). The nematomorphs are parasites as juveniles but free-living as adults (Poinar, 2010). In Tardigrada, females are iteroparous but males of some limnic species are semelparous (Bertolani, 2001). (viii) *Cells*: Known for constancy of cell numbers, the eutelic pseudocoelomates (e.g. gastrotrichs, Hummon, 1984; rotifers, Stelzer, 2005) are unique, as no further nuclear division occurs after hatching. In contrast, endogony is a cellular process, in which the myxozoan primary (sporoplasm) cell in its cytoplasm may generate and contain several hundred cells. By the same endogonic process, the entire gametogenesis in the 'gonad' is manifested within a single cell, the axioblast of Mesozoa. The eutely of pseudocoelomates and endogony of Mesozoa and Myxozoa open new avenues for fruitful cytological research.

In view of their evolutionary and economic (Rotifera and other parasitic phyla) importance, a few book series have been published on minor phyla. During the 1960s, Hyman (1951a, b, 1959) summarized the then available information on 16 minor phyla. Approaching the subject from an angle limited to structure and function of 22 minor phyla and other major phyla, Adiyodi and Adiyodi (1983, 1984, 1988, 1989, 1990, 1993, 1994, 1995, 2000, 2001) elevated the subject to a visible and recognizable status. Likewise, the series by Giese and Pearse (1974, 1975a, b, 1991) is limited to marine minor phyla. The book by Thorp and Roger (2014) is limited to freshwater phyla. There are other series (e.g. Woo, 2006, Saari et al., 2019) but they are limited to diseases caused by parasitic minor phyla. Fortunately, the new series by Wanninger (2015a, b, c) not only updated relevant information but also specifically introduced a section on molecular biology in each chapter. However, these multi-volume series are either taxon-based or habitat-based. For specific and/or comparative information on reproduction and development in minor phyla, one has to search different volumes. Briefly, there is a need for a comprehensive synthesis of relevant information on reproduction and development of all minor phyla in a single volume.

Part A
Aorganomorpha

Aorganomorpha, a new term to indicate the lack of organ in line with Acoelomorpha with no coelom, includes Placozoa (3+ species), Mesozoa (150 species) and Myxozoa (2,200 species) and are characterized by the presence of a few cell types, 5 in Placozoa and 7–9 in Myxozoa. They are gonochores and their parasitic life cycle is indirect, except in the free-living Placozoa. The incorporation of highly motile vertebrates as intermediate hosts has facilitated evolution and speciation in Myxozoa. The predominance of parasitism seems to have secondarily reduced most of them to aorganomorphism.

2

Placozoa

Introduction

The Placozoa are multicellular animals that lack symmetry and any organ. They are closest to the very root of metazoan origin and one of the key phyla for unraveling early metazoan evolution. Thanks to an impressive series of publications by B. Schierwater and his colleagues, interest on the enigmatic Placozoa has been revived after its discovery by Schulze in 1883. Presently, the Placozoa include a single species *Trichoplax adhaerens*. However, recent analysis of placozoan specimens from different oceans has revealed the presence of many cryptic species. Hence, it has the potential for biodiversity of several dozens of genetically, embryologically and ecologically identifiable species. They are marine, occurring in coastal waters up to 20 m depth and at temperatures between 10°C and 32°C (Eitel et al., 2013).

Structural Features

The amoeboid irregular 'hairy plate' (Schierwater, 2005) Urmetazoan (Eitel et al., 2011), *Trichoplax adhaerens* measures 5 mm in diameter and < 20 μm in thickness (Schierwater and Eitel, 2015). It lacks an oral-oboral axis and hence polarity (Fig. 2.1A), any organ, nerve and muscle cells, basal lamina and extracellular matrix (Schierwater, 2013). It has only four-five defined somatic cell types: (i) lower epithelial cell and upper epithelial cell, (ii) gland cell, (iii) fiber cell and (iv) small potentially 'omnipotent' cells (Jakob et al., 2014, Guidi et al., 2011). The fiber cells are sandwiched between the 'nutritive' lower epithelial and gland cells at the bottom and 'protective' upper epithelial cells at the top (Fig. 2.1B). The omnipotent marginal cells are relegated to the sides between the upper and lower epithelial layers (Schierwater and Eitel, 2015).

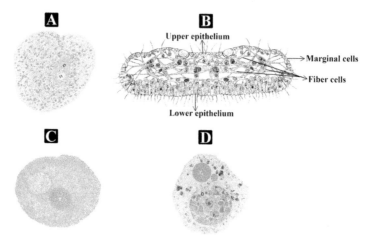

FIGURE 2.1

A. *Trichoplax adhaerens* (from Eitel et al., 2011, 2013). B. Schematic cross section of a placozoan to show the locations of the upper and lower epithelia, marginal and fiber cells (modified and redrawn from Eitel et al., 2013). C. A parent with one oocyte and D. animal with ovum and yolk (from Eitel et al., 2011). O = oocyte, N = nucleus, Y = yolk.

Motility and Feeding

Attached to the substratum by the lower epithelium, the animal is able to crawl aided by the cilia. Lifting the center region of its body, it is capable of forming an external digestive cavity between the substratum and lower epithelium (Schierwater, 2013). Algae and other food particles trapped in the slimy upper epithelium can also be phagocytized by the inner fiber cells. The animal may also engage endosymbiotic microbes in the endoplasmic reticulum of the fiber cells (Eitel et al., 2011). With the absence of chlorophyll but with symbiotic microbes, *T. adhaerens* is a herbivore. The microbes may synthesize only those nutrients, which the animal is not able to incorporate.

Reproduction

The placozoans are known to reproduce clonally in fields but both clonally and sexually in the laboratory. In three placozoan species lineages namely (i) the GRELL clone of *Trichoplax adhaerens*, (ii) the Placozoa sp 2, CAR-PAN-4' clone and (iii) Placozoa sp H16, KEN-A' clone, Eitel et al. (2011) have provided molecular evidence for sexual events for allele shuffling during sexual reproduction. They have brought to light new features of sexual reproduction, which are briefly summarized.

1. Oogenesis: Reared at high densities together with food scarcity, temperature (23°C) induces oocyte production from the lower epithelium within 4–6 weeks. Oocytes begin to appear at 23°C and above but not at 22°C, indicating a sharp thermal barrier between clonal and sexual reproduction. Over 70% 'female' animals develop a single oocyte each (Fig. 2.1C) that always gives rise to an embryo. Those with more than two oocytes, the extra ooctye(s) are usually resorbed. However, a few develop successfully up to three oocytes each. Nursed by fiber cells and incorporation of lipid droplets, the oocyte grows to 70–120 μm size. Notably, the lipid droplets are found inside the oocyte, while the yolk is accumulated in the nursing cells of the mother animal (Fig. 2.1A). Cortical granules, the key elements known to build cortex to afford protection and prevent polyspermy across different metazoan phyla, are evenly dispersed throughout the placozoan oocytes. Following fertilization, the granules disappear and a two-layered fertilization membrane is formed (Fig. 2.1D). Incidentally, the fertilization process is not described. The growth and maturation of the occyte is accompanied by degeneration of the mother animal. The degeneration leads to a round-shaped animal harboring the oocyte in the center. In *T. adhaerens*, sex ratio may be in the range of 0.8 'female': 0.2 'male'.

2. Spermiogenesis: Though the classical cytological process of spermatogenesis has not been described, the transcriptome analysis of the three placozoan species has provided molecular hints for the existence of spermatogenesis. Using whole genome and EST sequences as well as DNA cloning, five conserved sperm markers have been identified. They are (i) Mns1 and (ii) Meig1 controlling spermatocyte meiosis and spermatogenesis, respectively, (iii) Nme5 protecting oxidative stress during spermatogenesis, (iv) protein Dnajb13 associated with sperm flagellum and (v) protein Spag8 involved in sperm-egg recognition. Briefly, these markers cover spermatogenesis from early meiosis to sperm maturation with functional flagella and to sperm-egg recognition during fertilization. Interestingly, Spag8, Mns1 and Meig1 are homologous proteins of other taxa. The other two proteins Dnajb13 and Nme5 are homologs to those of mice that are known to play an important role in spermatogenesis of the mouse.

3. Development: Early embryo undergoes development by cleavage, which is total and equal from zygote to 128 cell stages. It grows up to 128-cell stage inside the mother animal, when the latter completely degenerates. However, it fails to grow beyond 250 cell stage in the laboratory.

4. Clonal reproduction: Two modes of clonal reproduction are known (Schierwater, 2005): 1. By fission, the animal grows to a certain size and

divides into approximately equal daughter offspring. The fission mode of clonal propagation can go on *ad infinitum*. Under unfavorable conditions, free-floating small spherical swarmers are produced. Molecular markers like *vasa* and *piwi* may be employed to identify the germ cells and their locations in the animal. Research is also required to know (i) the approximate number of epithelial cells that undertake oogenesis, and the location, number and cell type that are primed for spermiogenesis, (ii) the threshold number of clonal cells required to produce a successful swarmer and (iii) the approximate number of cells that trigger clonal reproduction by fission.

3

Mesozoa

Introduction

The exclusively marine, polyphyletic (Pawlowski et al., 1996, Schiffer et al., 2018) Mesozoa or Rhombozoa include ~ 150 species in 8 genera (GUWS Medical, 2019) and 2 families namely Orthonectida and Dicyemida (42 species parasitic on 19 cephalopod species in Japan), which is subdivided into Conovyemidae and Dicyemidae (Furuya and Tsuneki, 2003). They are bilaterally symmetrical, extremely simple, tiny, vermiform endoparasites of a large number of marine invertebrates like echinoderms, mollusks and annelids. They lack body cavity as well as differentiated respiratory, digestive, circulatory and nervous systems. Dicyemid mesozoans are unique for the existence of gonads and manifestation of gametogenesis and self-fertilization—all within a single axial cell. The uniqueness calls for the attention of cytologists.

Dicyemida

As a sac filled with reproductive elements, a dicyemid has only < 50 cells differentiated into 6–7 cell types. As parasitic adults, they exist in clonal nematogen and sexual rhombogen forms. They have ~ 42 cells: the rhombogen consists of (i) a central cylindrical axial cell, the axoblast and (ii) an outer layer of 8–30 ciliated peripheral cells (Fig. 3.1A). The number of peripheral cells is fixed and species specific. At the anterior region, 4–10 peripheral cells form the calotte, the cilia of which are shorter and denser than the more posterior diapolar peripheral cells (GUWS Medical, 2019). The calotte comprises of two tiers of the upper propolar and lower metapolar cells. In all, not more than six somatic cell types are present. With increasing body size, the number of peripheral cell increases; hence, the somatic peripheral cells undergo mitotic proliferation to achieve body growth (Furuya and Tsuneki, 2003). Incidentally, the infusoriform embryo also consists of ~ 7 cell

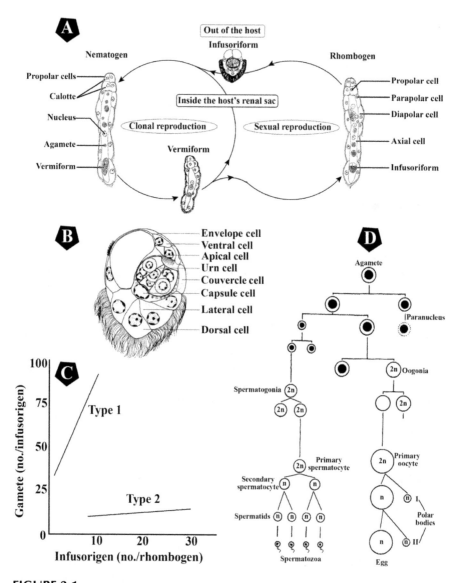

FIGURE 3.1

A. Clonal and sexual reproductive cycles of dicyemid within the host's renal sac. Note the freely released infusiform embryo out the host. B. Free hand drawing of infusiform embryo of *Dicyema acuticephalum* (ventral view). C. Relation between gametes (no./infusorigen) and infusorigens (no./rhombogen) in two distinct types of dicyemids (modified and redrawn from Furuya and Tsuneki, 2003). D. Gametogenesis within a single axial cell of *D. japonicum* (modified and redrawn from Furuya et al., 1996).

types (Fig. 3.1B). However, these cell types differ from those of rhombogen. Presumably, the larval cells are redifferentiated into those of adults.

Vermiform embryos of most species have a constant number of peripheral cells, except in *Dicyemia acuticephalum*, *D. benthoctopi* and *D. bilobum* (see Furuya et al., 1996). From its 'agamete', the nematogen clonally generates vermiform embryos; however, their dispersal and infection are restricted to the renal sac of the same host. High density of nematogen within a host may induce the vermiform embryo to develop into rhombogen (instead of nematogen, Lapan and Morowitz, 1972). Within a rhombogen, the axial cell serves as a hermaphroditic gonad and generates gametes. Details of gametogenesis in *Dicyemia japonicum* are illustrated in Fig. 3.1D. The 'agamete' undergoes an unequal division; the smaller daughter cell loses its cytoplasm and is named paranucleus, which does not participate in gametogenesis. The infusoriform embryo appears from a self fertilized egg and its larva consists of 37 cells. The process of development of vermiform embryo is simple and programmed to that of infusoriform embryo. In contrast to the size of vermiform embryos, a positive correlation between the infusoriform embryos and host body size suggests that the infusoriform embryo size is adapted to the host size. A large rhombogen may harbor 20 infusoriform embryos. Consequently, the available cytoplasm space within a single axial cell critically limits the extent of gametogenetic process. As a consequence, the correlation between the number of gametes per infusorigen and the number of infusorigen/rhombogen is positive, but it falls at different levels in different dicyemid species (Fig. 3.1C). From the correlations, two distinct types are recognized. Type 1 includes a large number of infusorigens from a smaller number of gametes and type 2 includes a small number of infusorigens from a large number of gametes. In type 1, many oocytes may remain unfertilized for want of an adequate number of sperms. In *Dicyema sullivani*, the number of sperms may even be smaller than the number of oocytes (McConnaughey, 1993). At an individual level in dicyemids, fecundity of a single rhombogen is not high, in comparison to other endoparasitic taxa. However, the clonal reproduction in nematogen compensates to bring parity with other endoparasitic taxa at the population level.

From fully documented cell lineage studies (Furuya et al., 1996, 2001), it was established that the lineages were highly conserved in vermiform embryos of dicyemids. The species characteristics appear in the later embryogenesis and are related to morphological evolution and speciation (Furuya et al., 2001). As dicyemids share a common feature, the cleavage is briefly summarized in infusoriform embryos of *D. japonicum*: It is holoblastic and synchronized up to the third one. The first cleave is meridional and followed by latitudinal. The third one results in the animal pole with four blastomeres and a vegetative pole with the other four blastomeres. Hence, polarity is established at the third cleavage. Notably, these blastomeres are spirally distributed, a feature reminiscent of those of flatworms. However, with subsequent asynchronized and unequal fourth division in 2a, 2b, 2c and

2d blastomeres, a 20 blastomere-stage is achieved. Beyond the 20-cell stage, the cleavage is not spiral but bilateral and thereby the bilateral symmetry is established (Furuya et al., 1996). Within the 4th and 8th divisions, progenitors of all cell types of are established (Table 3.1).

TABLE 3.1

Early cleavages that prime the cell types in dicyemid embryos (condensed from Furuya et al., 1992)

Division (no.)	Progenitor of Cell Types
4	Caudal cells to be placed at different locations, couvercle cell and germinal cells
5	Lateral and apical cells
6	Capsule cells, dorsal and lateral cells
7	Antero-lateral, ventral and enveloping cells
8	Internal cells and uron cells

Orthonectida

Within Mesozoa, parasitism has reduced the dicyemids to a greater extent than that in orthonectids. Consequently, the dicyemids have < 50 cells with a half dozen cell types, whereas the orthonectids have a few hundred cells including 10 nerve cells (Schiffer et al., 2018) and genuine outer circular and inner longitudinal (Sliusarev, 2008) contractile myofibrils (Kozloff, 1965). The vermiform body is covered by an outer 'jacket' consisting of tiers of ring cells (Fig. 3.2A): of the ring cells, some are adorned with cilia. Glycogen and mitochondria are abundant in the ciliated cells. For details on distribution of cilia, Kozloff (1971) may be consulted. Sliusarev (2008) indicated that interdigitating ciliary cells constitute a single row of epithelial layer with underlying intercellular matrix (Fig. 3.2B). In orthonectids, sexes are separate and distinguishable by size and number of rings. For example, *Ciliocincta subellariae*, an endoparasite on polychaete *Sabellaria cementarium*, the female is larger (265 µ in length, 23 µ in width, 38 or 39 tiers of ring cell) than the male (~ 135 µ in length, 21 µ in width, 29 tiers of ring cell). Testis is fairly long and extends from the level of 17th ring to the 29th ring. Sperm with a head of 1 µ in diameter and a flagellum of 5 µ in length is observed. Within the jacket, the body is filled with 11 to 12 oocytes (Kozloff, 1971).

Based on the descriptions of Caullery and Lavallee (1908a, b, 1910), the life cycle of *Rhompalura ophiocomae* is summarized by McConnaughey (1989, 1993). Accordingly, the short lived, non-feeding males and females are released from infected hosts (Fig. 3.2A). Following impregnation, the zygote develops into a ciliated larva, which escapes from the female. The larva gives rise to an amoeboid multinucleate plasmodium. A plasmodium, enveloped

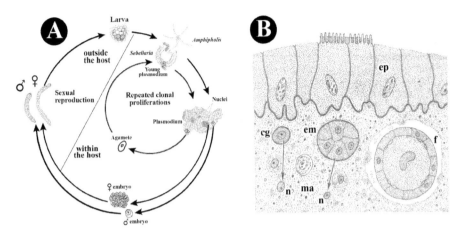

FIGURE 3.2

A. Sexual reproductive (outside the host) and clonal reproductive (inside the host) cycles of an orthonectid. B. Schematic diagram to show *Ciliocincta sabellariae* in the epidermis of *Sabellaria cementarium*. ep = epidermal cells, ma = matrix, cg = germinal cell, em = embryo, f = adult female orthonectid (in transverse section), n = nucleus arising from either germinal cells or embryo (drawn from Kozloff, 1997).

by two membranes (Sliusarev, 2008), is defined as a mass of (syncytium) cytoplasm with a large number of nuclei (Kozloff, 1997). A young plasmodium consists of cytoplasm, in which agametes or germinal cells are embedded. The germinal cells clonally multiply and the plasmodial progenies grow and ramify the host tissues. From isolated germ cells of the plasmodium, male embryos are formed, while female embryos appear from the transformation of moruloid groups of cells as a whole into embryos. Figure 3.2B shows the infected surviving epidermal cells and underlying matrix harboring adults, embryos and germinal cells.

Reports on the presence of plasmodial nuclei differ from author to author. Caullery and Lavallee (1908a, b) were perhaps the first to describe the presence of multinucleate plasmodium. From his electron microscopic study, Kozloff (1997) found no evidence for the presence of plasmodial nuclei. Investigations involving both intra- and inter-specific grafts led Haloti (1993) to conclude that the germinal cells are the pioneers that pervade the tissues of the recipient. In fact, he avoided using the term plasmodium. Reporting new data on morphology and biology of orthonectid plasmodium, Sliusarev (2008) confirmed not only the presence of plasmodium but also indicated that the reproductive individuals move within it. Interestingly, he also hints that the orthonectid life cycle includes alternative sexual, clonal and parthenogenic reproductive phases. He characterizes it as 'metaheterogony', a combination of the terms metagenesis and heterogony. It is not possible to provide more information on his contribution, as his publication (in Russian language) could not be obtained.

4

Myxozoa

Introduction

The 2,200 speciose polyphyletic (Gruhl, 2015) Myxozoa (in 62 genera, Lom and Dykova, 2006) are highly specialized endoparasites. If parasitism has reduced the dicyemid mesozoans to < 50 cells and a half dozen cell types, it has reduced myxozoa into the most simplified form of multicellular spores. Rarely present in other animals, the spore is omnipresent and a characteristic of myxozoans (Canning and Okamura, 2004). Most Myxozoa lack any form of alimentary, nervous and reproductive system (Gruhl, 2015). The lack of developmental genes (e.g. *Hox*-like *Wnt, Runx, Hedgehog*) suggests that the genes are no longer required in the highly simplified myxozoans (e.g. *Kudo iwati, Myxobolus cerebralis,* see Okamura and Gruhl, 2016). In a few malacosporeans (e.g. *Buddenbrokia plumatellae,* see Canning and Okamura, 2004), the presence of (i) a bonafide epithelium (including basal lamina, Gruhl and Okamura, 2015), (ii) four longitudinal muscle blocks and (iii) rows of connective cells (Fig. 4.1A) identify them as degenerated animals. A further step in degeneration is the loss of muscle blocks in *Tetracapsula bryozoides.* Yet, their life cycle is complex and involves a definitive oligochaete/bryozoan host and a vertebrate (mostly fish) intermediate host. With advent of aquaculture,

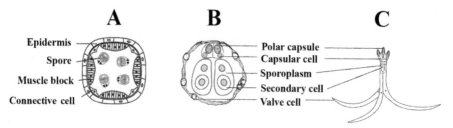

FIGURE 4.1

Schematic free hand drawing to show A. a section through malacsporean *Buddenbrockia* with four longitudinal muscle blocks and four rows of connective cells, B. malacospore and C. actinospore (free hand drawings from Gruhl, 2015, Lom and Dykova, 2006, Kallert, 2006).

the Proliferative Kidney Disease (PKD), whirling disease (see Eszterbauer et al., 2015) and ceratomyxosis in salmonids, the myxozoans have received much attention in recent years (e.g. Doctoral Dissertation: Kallert, 2006; review: Feist and Longshaw, 2006, Okamura et al., 2015).

Life Cycles

The myxozoans are classified into two classes: Myxosporea (or Actinosporea, 58 genera) and Malacosporea (2 genera). As definitive hosts, the former utilizes oligochaetes, while the latter engages bryozoans. The microscopic spores accomplish transmission between the two hosts. Though diverse in shapes (Lom and Dykova, 2006), their morphology is uniform. Remarkably, the vertebrate host phase of the myxozoan life cycle is dominated by endogonic multiplication resulting in a single spore cell containing one to several hundred cells in its cytoplasm. Uptake of nutrients is facilitated by endocytosis (Gruhl, 2015). Though parts of life cycle have been inferred for 100 (Gruhl, 2015) or 50 species, the complete life cycle has been resolved for five species only, four from Myxosporea *Ceratonova shasta* (syn. *Ceratomyxa shasta*), *Myxobolus cerebralis*, *M. parviformis*, *M. pseudodispar* and one malacosporean species *Tetracapsuloides bryosalmonae* (Eszterbauer et al., 2015). Of the complicated cycles in the two classes, the simpler is that of malacosporeans. The description of myxosporean life cycles is based on Feist and Longshaw (2006), Feist et al. (2015) and Gruhl (2015).

Malacosporea

All the known malacosporeans exclusively infect the freshwater phylactolaematid bryozoans (Gruhl, 2015). In malacosporeans, the cycle commences with the entry of fishmalacospore into a bryozoan host through mucus cells of its skin epithelium, the uninucleated sporoplasm undergoes sporogenic multiplication to form the nucleated sporoplasms (Fig. 4.2 left). Subsequently, the sporoplasm produces an uninucleated multicellular cluster, leading to the formation of pansporocyst. A pansporocyst is a simple sac delimited by two to eight cells enclosed by an epithelium consisting of flat interconnected cells. A mature sac is filled with eight malacospores. In each malacospore, a wall is formed by capsulogenic and flattened valve cells. Typically, it contains sporoplasm with a primary cell and a secondary cell (Fig. 4.1B). Prior to the formation of a pansporocyst, meiosis is assumed to occur (Canning et al., 2000). According to Canning and Okamura (2004), malacospores are derived from 'Type β cells' that meiotically divide by binary fission to form a cluster of 14 cells destined to differentiate into malacospores.

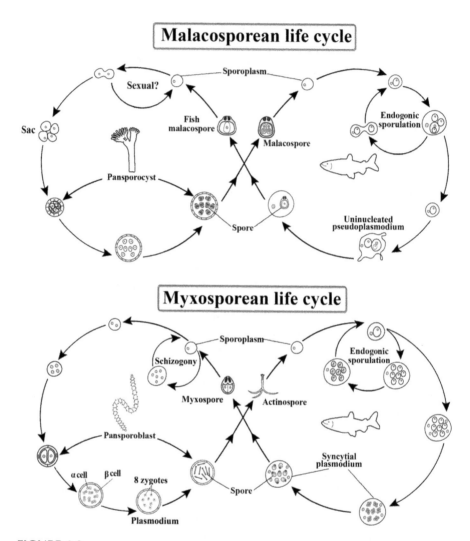

FIGURE 4.2

Life cycles of malacosporean *Tetracapsuloides bryosalmonae* in definitive bryozoan host (left) and intermediate fish host (right) and myxosporean (e.g. *Myxobolus cerebralis*) in definitive oligochaete host (left) and intermediate fish host (right). Note the presence of α and β cells within pansporoblast of myxosporean and non-sexual endogonic sporulation cycles within fish host (modified and compiled free hand drawings from Feist and Longshaw, 2006, Kallert, 2006, Lom and Dykova, 2006, Gruhl, 2015).

Following the release of *Tetracapsuloides bryosalmonae* malacospore from the pansporocyst in the bryozoan host, the binucleated germinal sporoplasm, on successful infection, enters the fish through its mucus cells (Fig. 4.2 right). However, the sporoplasm becomes cryptic/dormant for different durations.

Triggered by the host condition, it undergoes endogonic sporulation by proliferation through open mitotic (Morris and Adams, 2008) divisions rapidly into sacs or clusters, each consisting of many to several cells with sporogenic cells floating freely in the inner cavity. Depending on developmental stage and species, within each of this sporoplasmic primary cell, secondary, tertiary and even quaternary spores are produced by endogonic sporulation, resulting in the production of a uninucleated cell, the pseudoplasmodium (Gruhl, 2015). According to McGruk et al. (2005), each spore of *T. bryosalmonae* is ~ 19 µm in diameter with eight valves enclosing two putatively haploid, sporoplasmic cells and capsulogenic cells, each containing the spherical polar capsule with internalized coiled polar filament. From the pseudoplasmodium, fish malacospores are produced.

Myxosporea

The myxosporean life cycle (e.g. *Myxobolus cerebralis*) is more complicated than that of malacosporean, especially with regard to 'sexual' cycle within the oligochaete host and formation of myxospore or actinospore. According to Feist and Longshaw (2006), a uni- or bi-nucleate sporoplasm is released from myxospore after the ingestion of decaying infected fish by an oligochaete worm (Fig. 4.2 left). On contact with the gut epithelium, the polar capsules discharge and release amoeboid binucleate sporoplasm. Following gut penetration, the sporoplasm undergoes presporogenic (schizogonic) proliferation resulting in the multinucleated stage. After this plastomy, a 4-nucleated sac is formed. Of them, the more peripheral two cells enclose the other two cells. The outer enveloping cells undergo mitotic divisions to produce a pansporoblast wall containing of 2–16 diploid cells. Subsequent to meiotic divisions, 16 haploids, i.e. 8 haploids for each of the 2 cells, are produced. A zygote is produced by the fusion of α and β cells. It undergoes two mitotic divisions to produce a four-cell stage. The next phase is sporogony; three of the more peripheral cells in the four cells, mitotically divide to produce six cells in actinospores with three valves. Of them, three cells become capsulogenic and the other three valvogenic cells. The fourth in the 4-cells becomes a sporoplasm cell, which by an internal cleavage differentiates into a generative cell surrounded by a somatic cell. Some myxosporeans may have 2 or even 12 sporoplasm cells (e.g. *Polysporoplasma*). In fish, the cycle commences with actinospore, which is formed in the plasmodium of the oligochaete host. An actinospore consists of three uninucleated valve cells at the bottom, three capsular cells at the top, and the middle nucleated secondary cells within the sporoplasm (Fig. 4.1C). From simultaneous differentiation within pansporoblast, valvogenic, capsulogenic and sporoplamagenic cells are formed. The valvogenic cells are then attenuated to encircle the cytoplasm and capsulogenic cells and subsequently disintegrate. Capsulogenic cells, abundant with endoplasmic reticulum, differentiate into capsules and are internalized within the tube. On maturation, each capsular cell contains three coiled filaments. The extrusion of the filament releases the sporoplasm. Most actinospores are adorned with characteristic ridges, bristles, pits, mucous envelops (to resist a host immune response) or a variety of valvular extensions, all of which play a role in the distribution and migration of the parasites.

On contact with a suitable intermediate fish host, the polar filaments are discharged from the actinospore to anchor it to the host (Fig. 4.2 right). The favorable entry site is through mucus cells. Following degeneration of the cell wall surrounding the actinospore, the sporoplasm pervades surrounding tissues. In fish hosts, further development by endogonic sporulation is species specific and varies between genera. It ranges from production of small, uninucleate psuedoplasmodia to large, complex extrasporogenic proliferation stages prior to spore formation. For more details, Feist and Longshaw (2006) may be consulted. According to Gruhl (2015), endogony generates one or few spore(s) within a small pseudoplasmodium or more spores within the large plasmodium. Two pathways are common. In the plasmodium, the spores are individually formed and the secondary cells undergo repeated mitotic divisions to form sporoplasms, valves and capsulogenic cells. Or large populations of secondary cells are produced, which separately develop into sporoplasms, valve cells and capsular cells. Subsequently, these cells aggregate to form the myxospore.

Cell and Cell Types

Myxozoans are multicellular aorganomorphic animals. A single primary sporoplasmic cell may generate and contain one to several hundred cells within its cytoplasm (Feist et al., 2015). Fishmalacospore, malacospore, myxospore and actinospore are all multicellular organelles of myxozoans. Pansporocyst is a multicellular stage but the syncytial plasmodium and uninucleated pseudoplasmodium are unicellular stages in the life history. Hence, (i) valve cell, (ii) capsular cell, (iii) sporoplasm, (iv) plasmodium, (v) pseudoplasmodium, (vi) 'gamete' cell and (vii) epithelial cell enveloping pansporocyst are the common cell types. Other than these seven cell types, some myxozoans like *Buddenbrokia plumatellae* possess the reminiscent (viii) muscle block cell and (ix) the connective cell. In all, the myxozoans are composed of not more than nine cell types. *Polypodium hydriforme*, a parasitic schiphozoan and close relative of myxozoans, corresponding in fine structure—the myxozoan polar capsule and cnidarian nematocyst—

and protein composition, has also ~ seven cell types ((a) ectodermal cells, (b) tentacular cells, (c) cnidoblasts, (d) endodermal cells, (e) glandular cells, (f) muscle cells, (g) nerve cells and (h) germ layer cells). Parasitism has reduced systems, organs, and tissue types but it cannot reduce the tissue types to less than a critical minimum of 7–9 cell types in these parasitic diploplastic medusa *P. hydriforme* and aorganomorphic parasites like Mesozoa and Myxozoa.

Prevalence and Distribution

Information on myxozoan prevalence is limited. But the available few publications seem to indicate that the prevalence may serve as an indicator of the transmission efficiency of malaco- and myxo-spores. Beauchamp et al. (2002) reported a mean prevalence of 8.3% (range 4.2 to 14.1%) of the myxosporean *Myxobolus cerebralis* on *Tubifex tubifex*. From a second field study, they reported that of 7,725 *T. tubifex*, 411 were found infected, i.e. 0.6% prevalence. In a laboratory experiment, they also exposed 5,421 *T. tubifex* to infection and found a prevalence of 7.6%. Thirty-one samples of *Tetracapsula bryozoides* collected from 18 lakes on different dates revealed no infection of malacosporean *Cristallata mucedo* in 8 samples containing 174 specimens, i.e. 39.4% bryozoans were not infected. Of the remaining 1,222 specimens, prevalence ranged from 2.6% in backwaters of the Thames River to 49.2% in the Beale Bird Park Lake. Understandably, the prevalence varied temporally and spatially. On the whole, 16.2% of the sampled bryozoans were infected (Okamura, 1996). These publications hint that the transmission efficiency of malacospore is higher (16.2%) than that of ~ 7.6–8.3% for myxospore. More specifically, the trophic mode of infection of the oligochaete worms by myxospores seems to be less efficient than the contact mode of infection by malacospores on the relatively larger surface area of the lophophorate bryozoans.

With incorporation of highly motile vertebrates as the intermediate host, the prevalence of myxozoans increased remarkably. In the myxozoan *Parvicapsula minicicornis*, it ranges from 20–27% to 100%; the highest prevalence occurs at the spawning site, where the salmon density is the highest (Jones et al., 2003). Clearly, host density is another important factor in the determination of prevalence. An experimental study by Kallert (2006) indicated that *Henneguya nuesslini* successfully infected a couple of salmonid species but not *Cyprinus carpio*. Thanks to the flourishing aquaculture, a series of Indian publications have reported the incidence of ~ 56 myxosporean species on a dozen Indian major carp species as well as on *Channa punctatus*, *Nandus nandus*, *Mastacembilus armatus*, *Tor putitora* and *Wallago attu* but not on *Ch. striatus*, *Chitala chitala*, *Eutropiichthys vacha* and *Trichogaster* sp.

Susceptibility of these fish host species decreases in the following descending order: *Labeo rohita* (18 parasitic species) > *Cirrhinus mrigala* (14 species) > *Catla catla* (12 species) > *L. calbasu* (11 species) > *Ci. reba* (8 species). These polyspecies parasitism reveals that the existing myxosporean species do not inhibit the infection by one or more incoming myxosporean species (see also Ahmad and Kaur, 2018). The preferred sites of infection also decreases in the following descending order: gills (45.6%) > caudal fin (29.4%) > scales (16.1%) > duodenum (5.1%) > pectoral fin (1.9%) > pelvic fin (1.7%) > eye ball (1.4%) > snout skin (0.4%) > stomach (0.1%) (Kaur, 2014). The prevalence in some of these fish ranges from 13% for *Myxobolus chirsurahensis* in *C. catla* to 52% for a combination of *M. karnaticus*, *M. orissae* and *Thelohanellus caudatus* (Table 4.1). Hence, a combination of highly motile intermediate Indian major carps and high densities in polyculture system has facilitated higher prevalence of myxosporeans.

TABLE 4.1

Prevalence of myxosporeans on intermediate host

Myxozoan Parasite	Fish Host	Prevalence (%)	Reference
Henneguya mystusia	*Mystus vittatus*	25.5	Kaur (2014)
Myxobolus, Thelohanellus, Henneguya	*Mystus vittatus* Labeo bata	47.6 11.1	Banerjee et al. (2016)
M. chinurahensis	*Catla catla*	13.4	Kaur (2014)
M. knobii, M. majraiensis, M. markiwi, M. naini, M. nanokiensis, M. potularis, M. rocatlae, M. slendrii, M. vascularis, M. venkateshi	*Carrasius carassius Catla catla, Cirrhinus mrigala, Labeo rohita*	13.7	Ahmad and Kaur (2018)
M. chirsurahensis	*Catla catla*	13.4	Kaur (2014)
T. auerbachi	*L. bata*	19.5	Kaur (2014)
T. catlae	*Puntius phutunio*	21.5	Kaur (2014)
T. bifurcata	*Salmostoma bacalia*	26.2	Kaur (2014)
M. oliveirai	*Brycon hilarii*	38.1	Milanin et al. (2010)
M. karnaticus, M. orissae, T. caudatus	Indian major carps	51.6	Ramudu and Dash (2016)

Myxozoans are common parasites of teleosts, elasmobranchs (Heupel and Bennet, 1996) as well as aquatic amphibians and reptiles. Some of them are highly host specific, while others are specific to a family (e.g. *M. cerebralis*, *Tetracapsuloides bryosalmonae*). Still others are generalists (e.g. *M. aeglefini*) and infect both gadoids and pleuronectids. In general, the definitive oligochaete hosts release spores throughout the year, but only during spring and summer in temperate species. This is also true of bryozoans, for example, *T. bryosalmonae* spawns only during spring and summer. Extrasporogenic stages of some like *T. bryosalmonae* are short-lived but others live long, potentially persisting for the life time of the host (Feist and Longshaw, 2006). Using the naturally occurring parasites (e.g. Trematoda, see Pandian, 2020), the stock structure of marine fish is investigated. The myxosporeans *Myxidium oviforme* and *Zschokkella hildae* are shown to be suitable tags for the stock assessment of *Gadus morhua* (Feist and Longshaw, 2006).

Part B
Acoelomorpha

Acoelomorphs mark the development of organs and systems. They include four minor phyla and one major phylum, the Platyhelminthes. The former are gonochores and may involve an indirect life cycle with the exception of Gnathostomulida. In contrast, platyhelminths are hermaphrodites (except for a few schistostomes) and involve an indirect life cycle with one or more intermediate host(s). In them, sexual reproduction, an indirect life cycle and incorporation of intermediate host(s) seem to have facilitated evolution and speciation (27,700 species, Pandian, 2020) of platyhelminths, whereas the direct life cycle and consequent reduction in lifetime fecundity (1–6 eggs) have limited the scope for evolution and speciation among the minor phyletic acoelomorphs.

5

Loricifera

Introduction

The monophyletic loriciferans are microscopic (80–800 μm in length) and exclusively marine, interstitial acoelomates; they are bilaterally symmetrical, with spiny heads retractable into abdominal cuticularized lorica (Neves et al., 2014). It was Kristensen (1983), who first erected the phylum with a description of *Nanaloricus mysticus*. Since then 34 species have been described and > 50 nominal species remain to be described. The known loriciferans are placed in 11 genera and 3 families: Family Nanaloricidae including *Armorloricus*, *Australoricus*, *Culexiregiloricus*, *Phoeniciloricus* and *Spinoloricus*, Family Piliciloricidae including *Piliciloricus*, *Rugiloricus* and *Titaniloricus*, and Family Urnaloricidae with *Urnaloricus*. The discovery of *Spinoloricus turbatio* and *C. trichiscalida* from deep anoxic hypersaline L'Atlantic Basins opens new avenue for a better understanding of the evolutionary history of Loricifera (Neves et al., 2014).

Sexuality and Life Cycle

The loriciferans are gonochores. The male is smaller than the female; it is covered with three pairs of clavoscalids, and has a pair of large dorsal testes. The female is covered with four pairs of clavoscalids and has a single ovary. The life cycle varies not only with species but also with feeding conditions in the same species. The fertilized egg develops into the Higgins larva (Fig. 5.1), which, following a few molts becomes a post-larva without gonad and toes. The larva molts again to become an adult. In *Rugiloricus carolinensis*, the Higgins larva molts directly to a mature male or female. Some Higgins larvae grow larger than the adults and develop a large single ovary. Subsequently, it enters into a cyst-like stage. Within the cyst, the larva molts repeatedly and generates 4 to 12 or more eggs, which develop into

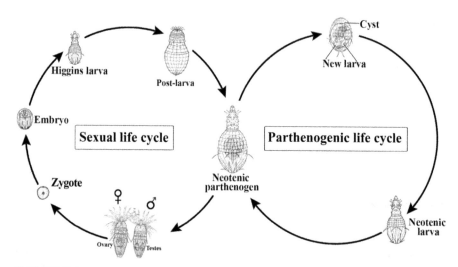

FIGURE 5.1

The proposed sexual and parthenogenic life cycles of the loriciferan *Rugiloricus*. Note: the neotenic parthenogen can either switch into sexual reproductive cycle or parthenogenic life cycle (modified and redrawn from Kristensen, 2002).

the so called neotenic larvae. The neotenic larva molts a number of times and repeats the neotenic parthenogenic cycle several times. Presumably, the neotenic cycle involves ameiotic parthenogenesis. However, no cytological evidence is yet available to show whether it is an ameiotic or endomitotic parthenogenesis.

6

Cycliophora

Introduction

The cycliophorans are bilaterally symmetrical, microscopic (350 μm in length, 100 μm in width), exclusively marine acoelomates. They are epizootically sessile on the setae of the mouth parts of European and American lobsters. With the description of *Symbion pandora*, Funch and Kristensen (1995) erected the new phylum Cycliophora. With subsequent addition of *S. americanus*, the phylum consists of only two species. In its predominantly largest feeding stage, the cycliophoran body, covered with thick well-differentiated cuticle, is divisible into an anterior buccal funnel with a mouth and esophagus, an oval shaped trunk and a short acellular stalk with an adhesive disk. In the trunk, the space between epithelial layers is totally occupied by large vacuolated mesenchymal cells, leaving no space for body cavity. The bands of compound cilia beat as a downstream collecting system, indicating that they belong to Protosomia (Funch and Kristensen, 1995). Internal anatomy of the trunk includes (i) a U-shaped gut terminating in an anus at the anterior end of the trunk (Kristensen, 2002), (ii) a set of circular, longitudinal, dorso-ventral and other muscles and (iii) bilobular ganglion with two ventral nerve chords, as well (Wanninger and Neves, 2015). The sessile feeding stage produces a single inner bud with the new buccal funnel and intestine. When the bud is fully developed, the old buccal funnel is sloughed off. This sort of molting recurs several times (Kristensen, 2002).

Life Cycle

The cycliophoran metagenic life cycle is complicated and consists of clonal and sexual generations (Fig. 6.1). The feeding stage clonally generates one at a time the Pandora or Primetheus larva and the sexual female. The Pandora emerges, when the buccal funnel is cast off. It is a poor swimmer and soon

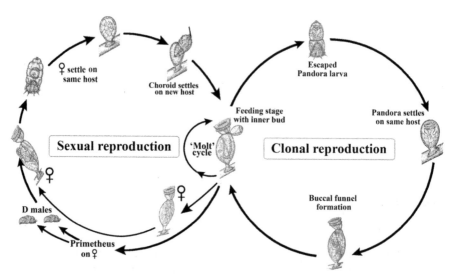

FIGURE 6.1

Sexual and clonal reproductive cycles in *Symbion pandora*. Note the largest feeding stage undergoes repeatedly partial 'molt' cycles. It clonally gives rise to Pandora larva, which after escaping from the parent reinfects the same host. The feeding stage may switch to generate sexual female, and Primetheus larva, which settles on a female and develops into two or three dwarf males. A fertilized female escapes from the feeding stage and develops into choroid larva, which settles on a new host (modified and redrawn from Kristensen, 2002).

settles with head glands close to the parent individual and gives rise to new feeding stage (Kristensen and Funch, 2001).

The other clonal part of the life cycle consists of two types of feeding animals, the first internally generates the Primetheus larva and the second type the sexual female (Kristensen, 2002). On its release, the Primetheus larva settles on the trunk of a feeding individual, encysts and subsequently generates one to three dwarf males, each with a cuticular penile structure (Wanninger and Neves, 2015). The escaping sexual female settles on the same host. Through a yet to be described mechanism, insemination occurs, following which embryonic development commences, while the female is still perching on the feeding stage. Subsequently, emerging from this female, the choroid larva settles on a new host.

The Dwarf Male

With a body length of ~ 40 µm and 50 nucleated somatic cells, the cycliophoran dwarf male is among the smallest free-living metazoan. Not surprisingly, the complex body plan of the minute, free-living cycliophoran male has attracted considerable interest. Some smallest known halothuroid *Panotrochus belyaevi* (Pandian, 2018) measures 1.5 µm in length and oelosomatid annelid weighs < 100 µg (Pandian, 2019). From an ultrastructural investigation, Neves and Reichert (2015) traced the differentiation and development of dwarf males from the Primetheus larva to adult of *S. pandora*. An attached Primetheus larva contains three dwarf males—the first one at a mature stage and the second at a maturing stage—and the third largely and undifferentiated bud-like stage. In the dwarf males, the differentiation and maturation involve the process of miniaturization by massive reduction in the number of nucleated somatic cells, especially epithelial and mesenchymal cells from the size of 13,900 to 9,000 µm³ (Table 6.1). After organogenesis, miniaturization occurs with the absence of cellular multiplication, programmed cell death and massive loss of cell nuclei. Miniaturization of the body cells is also known in some fishes like *Schindleria praematura* (Pandian, 2011).

TABLE 6.1

Differentiation of *Symbion pandora* dwarf male by miniaturization and massive reduction in cell number (compiled from Neves and Reichert, 2015)

Features	Mature ♂	Maturing ♂	Budding ♂
Body size			
Length (µm)	27	30	29
Volume (µm³)	9,000	14,300	13,900
Nucleated cell (no.)			
Epithelium/mesenchyme	3 (6.4%)	114 (55%)	-
Nervous system	34 (72%)	49 (23.6%)	-
Musculature	0	32 (15.4%)	-
Secretary glands	8 (39%)	8 (17%)	-
Testes	2 (4.3%)	4 (1.9%)	-
Prostate glands	-	1 (0.5%)	-
Total	47	208	186

7

Nemertea

Introduction

The monophyletic, 1,300 speciose (von Dohren, 2015) (usually named as Nemertina, Nemertini, Rhynchocoela) are acoelomic, cylindrical or flattened elongated eutelic worms with an anus. Their body bears no definite head but an anterior eversible proboscis is enclosed in a characteristic fluid-filled tubular cavity, the rhynchocoel located dorsal to the gut (Hyman, 1951a). The unsegmented body is housed within a monolayered, ciliated epidermis and contains several layers of muscles constituting the body wall, circulatory system with a pair of lateral vessels, pronephridial excretory system, and cerebral, frontal and lateral neural organs as well as sensory organs including the eyes and statocysts (Kolasa and Tyler, 2010). For more details on the anatomic features, Hyman (1951a) may be consulted. The smallest known nemertean is *Carcinonemertes* (Fig. 7.2G) and the longest *Lineus longissimus* (Fig. 7.2H); in fact, *L. longissimus* measures 60 m in length and is the lengthiest animal known on Earth (Goransson et al., 2019). Research on the chemical nature of toxic substances secreted by predatory nemerteans have led to the identification of several compounds of potential medical use or application in biotechnology (Goransson et al., 2019). Nemertean egg predators cause substantial losses to commercial crustacean fishery both in capture and aquaculture (Kuris, 1993).

Taxonomy and Distribution

Based on the number of body wall musculature and location of the mouth in relation to anlage of the brain, nemerteans are classified into two subclasses namely Anopla and Enopla (Fig. 7.1). The former is divided into two orders: Paleonemertini and Heteronemertini and the latter into Hoplonemertini and Bdellonemertini, which is further broken down into two suborders: Monostilifera and Polystilifera (Hyman, 1951a). On the basis of the presence

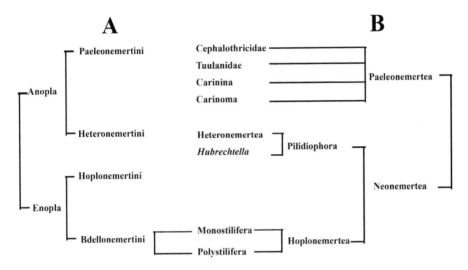

FIGURE 7.1

Classification of Nemertea according to A. Hyman (1951a) and B. Andrade et al. (2012, 2014) (compiled from Hyman, 1951a, von Dohren, 2015).

of mid-dorsal blood vessel (Neonemertea), pilidium larva (Pilidiophora) and proboscis equipped with stylet apparatus, Andrade et al. (2012, 2014) developed a consensus nemertean phylogeny (see also von Dohren, 2015). In Fig. 7.1, an attempt has been made to present a comparative picture of these two classifications. Of 1,300 described species, 1,265, 22 and 13 are marine, freshwater and terrestrial habitants, respectively (Goransson et al., 2019), suggesting that the nemerteans are perhaps the first invertebrates to venture and 'colonize' the *terra firma*. After drying, the terrestrial *Neonemertes* is provided with mucous and frontal glands. In *N. agricola*, the eggs develop directly into juveniles within an ovarian sac (Hyman, 1951a). Most marine nermerteans are benthic; some descend to greater depths from 200 m to 3,000 m depth; some others are intertidal (e.g. *Lineus ruber*, *L. viridis*); still others are pelagic (e.g. Polystilifera, *Neuronemertes*, Fig. 7.2K). The body shape of bathypelagic nemerteans is broad and flattened (Fig. 7.2I). The nemerteans are abundant in Arctic and Antarctic waters and many are circumpolar (e.g. *Parborlasia corrugatus*). However, they are less common in tropical and subtropical waters. Their cilia are paralyzed by lithium salts and gliding ceases, when their musculature is paralyzed by magnesium salts. These observations suggest that the nemerteans are capable of slow swimming and gliding (Hyman, 1951a). Nemerteans are either predators or scavengers. The level of their voracious feeding on decapod eggs has led some authors to suggest that they are parasites.

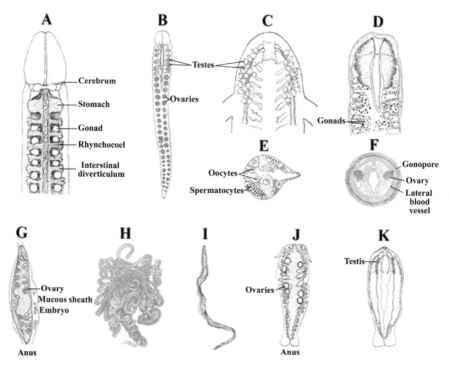

FIGURE 7.2

A. Schematic representation of internal structure of a gonochoric nemertean (freehand drawing from Riser, 1974), B. hermaphroditic *Dichronemertes* showing the testes and ovaries (after Coe, 1940), C. anterior part of *Nectonemertes* showing testes (after Coe and Ball, 1920), D. anterior part of parasitic *Gononemertes* with numbers of irregularly strewn gonads (after Brinkmann, 1927), E. ovotestis of *Prostoma rubrum* (after Coe, 1943), F. schematic transverse section through *Procarinina* (after Nawitzki, 1931), G. Viviparous *Carcinonemertes* within a mucous sheath (after Humes, 1942), H. longest nemertean *Lineus longissimus* (from Smithsonian website Frost, 2013, smithsonian.com), I. flattened body shape in pelagic haplonemertean *Drepanophorus crassus* (after Burger, 1895), J. *Balaenonemertes* female showing ovaries (after Brinkmann, 1917), K. *Neuronemertes* male showing testes (after Coe, 1926) (all figures, except G are redrawn from Hyman, 1951a).

Reproductive Biology

Sexuality: Most nemerteans are gonochores. However, hermaphrodism is reported mostly from freshwater and terrestrial nemertean genera: *Coenonemertes caravale, Dichonemertes, Poikilonemertes vivipara, Prostoma* (106 species), *Sacconemertella lutulenta* and *Geonemertes palaensis, Neonemertes agricola, Tetrastemma caeum, T. hermaphroditicum, T. kefersteinii* and *T. marioni*, as well (Riser, 1974). They are protandrous, except perhaps in *Prostoma rubrum*. An ovotestis is present in protandrous *P. rubrum* and *Procarinina* (Fig. 7.2E–F) and *S. lululenta*.

Gonadal development: For the gonad formation, some early events are described by Riser (1974). In *Lineus ruber*, Primordial Germ Cells (PGCs) first appear in the parenchyma (Olivier, 1966). As the PGCs enlarge, a membrane is formed surrounding the aggregate/sac. In the regenerating worm *Cerebratulus lacteus*, the gut wall grows outward between sacs. In *L. longifissus*, the ovaries, composed of a stroma of spindle-shaped cells, extend from the preformed oviduct and ramify into the connective tissue of the interdiverticular space and extend above, below and outside the diverticula. The development of testes is similar to the formation of the ovary.

The gonads are rounded or retort-shaped sacs that are limited to the intestinal region. They are located between the intestinal diverticula on each side mostly in singles or sometimes in groups (Fig. 7.2A, G, J, K). With the absence of the diverticula, they also occur in a regularly spaced row in the lower Palaeonemertini. In hermaphrodites, the testes are located in the anterior anlage and ovaries posteriorly (Fig. 7.2B). Numerous gonads are irregularly strewn throughout the mesenchyme in the so called parasites (e.g. *Gononemertes*, Fig. 7.2D) and commensals (e.g. *Nemertopsis actinophala*). In pelagic hoplonemertini, they are limited to cephalic (Fig. 7.2C) or the foregut region, where they occur in a paired row or cluster (Hyman, 1951a). Sex can be distinguished by cream ventral surface of the female *Amphiporus angulatus* from the red male; bright red intestinal region of *Cerebratulus lacteus* female from brownish red males (see Riser, 1974). In pelagic species, males have external papillae associated with each testis (e.g. *Pelagonemertes brinkmanni*).

Fecundity: For nemerteans, information available on egg size and fecundity is limited. The size ranges from as small as 50 µm in *Carinina romanei* to as large as 2.5 mm in *Dinonemertes investigatoris*. However, most values for the size fall between 100 and 300 µm in diameter. In this wide spectrum of egg size, the small eggs of paleonemerteans go to the lower end, while the larger eggs of hoplonemerteans (e.g. *Pantinonemertes agricola*, 350 µm–450 µm) to the other end of the spectrum (see von Dohren, 2015).

In low motile and patchily distributed animals, selfing provides a reproductive assurance. In selfers, inbreeding depression results from the expression of deleterious recessive alleles or from overdominance (e.g. Charlesworth and Willis, 2009). They may exhibit one of the two evolutionary syndromes: a combination of high inbreeding depression with low levels of selfing or low inbreeding depression with high levels of selfing. The limited available information on nemertean fecundity seems to indicate that the selfing in the simultaneous hermaphroditic nemertean *Prosorhochmus americanus* belongs to the combination of low inbreeding depression with high levels of selfing. Reared either in isolation (selfing alone) or a pair (crossing but with scope for selfing) for a period of ~ 160 days, fecundity, expressed in offspring number increases with increasing body length in both isolated and paired groups; however, the isolated worms are more fecund than those reared in pairs (Fig. 7.3A). From Fig. 7.3B, the following may be inferred: (i) The level for the surviving embryo number vs body size relation is higher for the isolated worms (covering a wider range of body size) than that for the paired worms. Hence, the higher level of selfing results only in low level of inbreeding depression. Incidentally, this observation counters the gradual decrease in offspring production attributed to the depletion of allosperm in seminal receptacles of simultaneous hermaphrodites maintained in isolation following mating event (e.g. turbellarians, Chintala and Kennedy, 1993, sea hares, Yusa, 1996). *P. americanus* lack seminal receptacles and cannot store allosperm after mating. Besides, with increasing number of mating partners, sperm competition is intensified and more number of sperms have to be produced at the cost of egg production in hermaphroditic *P. americanus* with an ovotestis, in which both eggs and sperm are produced within a single undelimited ovotestis (see Fig. 7.2E). In fish (Pandian, 2013) and higher invertebrates (e.g. Crustacea, Pandian, 2016) ovary and testis are accommodated in separate delimited chambers. The positive linear trends between offspring number and body size in *P. americanus* also confirms that the body size, by providing space to accommodate ripening eggs and developing embryos, is an important factor in determination of fecundity in oviparous and viviparous species (see also Pandian, 2016, 2017). (ii) With advancing age of *P. americanus*, the proportion of parental worms harboring embryos increased rapidly from ~ 20% at the age of ~ 3 weeks to 90% at the age of ~ 12 weeks and subsequently leveled off. This asymptotic level corresponds with that of the growth vs age relation (Fig. 7.3B). Clearly, body size is more important than age in determination of fecundity. (iii) *P. americanus* seems to produce embryos of different sizes. With increasing offspring size, the number of offspring decreased both in isolated and paired worms, albeit at different levels (Fig. 7.3C). It is an established fact that with

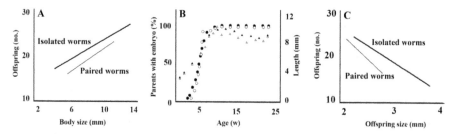

FIGURE 7.3

Prosorhochmus americanus: (A) Fecundity, expressed in offspring number, as a function of body size in isolated and paired worms. (B) Percentage of parents harboring embryos and growth as function of age in isolated and paired worms. Growth in isolated (▲) and paired (△) worms; Embryo number in isolated (●) and paired (○) groups. (C) Offspring number as function of offspring size in isolated and paired worms (modified, compiled and redrawn from Caplins and Turbeville, 2015).

increase in size of egg/embryo, the number of eggs/embryos shall decrease. The paired worms were unable to produce offspring larger than 2.5 mm size perhaps due to competition between the paired worms for food.

Gametes and fertilization: Free-living nemerteans aggregate prior to spawning. In some species, viscid mucous secretion around the worms reduce the space, into which eggs and sperms are released (e.g. *Carcinonemertes epialti*) and may facilitate internal fertilization (e.g. *Lineus ruber*, Bierne and Rue, 1979). In all others, fertilization is external. The fertilized eggs are laid in gelatinous strings and masses; in the jelly, eggs remain separate or grouped inside pyriform capsules. In *L. ruber*, eggs are laid in two types of string. Eggs in the green string are large and develop into viable larvae. Small numerous eggs in the red string die during development and are devoured by adelphophagic larvae arising from the green string (see Hyman, 1951a, Martin-Duran et al., 2015). It is not known how the oogenesis is altered to generate two types of eggs. A vast majority of temperate and polar nemerteans breed during summer (von Dohren, 2015).

Nemertean spermatozoa comprise a sperm head with an apical acrosomal vesicle, a condensed nucleus, a mitochondrial mass and diplosomal centrioles, to which a single flagellum is attached. In some hoplonemertean species, the sperm head is elongated due to an extended acrosomal vesicle. The elongation may facilitate the sperm penetration through vitelline layer and a fairly thick gelatinous mucous layer enclosing the egg proper (e.g. *Paranemertes preregrina*, see von Dohren, 2015). Self-fertilization may occur in hermaphroditic species (Hyman, 1951a).

Embryonic development: In his very impressive chapter, von Dohren (2015) has summarized available information on embryonic and larval development of nemerteans. In the relatively uniform embryonic development, cleavage is holoblastic, equal, spiral and determinate type. A notable exception is that of *Lineus ruber*, in which the first cleave results in an unequal division with a large CD blastomere and smaller AB blastomere (Nusbaum and Oxner, 1913). As a result, experimental investigations have revealed a range of blastomere potency. Halved at 2-cell stage, the plilidiophoran *Cerebratulus lacteus* embryo is capable of developing completely normal but miniaturized two larvae (Horstadius, 1937). In contrast any blastomere from a 2-cell or 4-cell stage develops into deficient larva in the hoplonemertean *Nemertopsis bivittata* (Martindale and Henry, 1995). In the spiralian development, mesoderm is derived from two sources: (i) the ectomesoderm derived from micromeres of either the second or third quartet and (ii) endomesoderm arising from the fourth quartet, the 4-Day mesendoblast (see von Dohren, 2015). Notably, the mesoderm arises without the ectomesodermal contribution in *Lineus ruber* (Nusbaum and Oxner, 1913).

Gastrulae and larvae: Subsequent to gastrulation, the development follows three different patterns: (1) In direct development, the oval or spherical

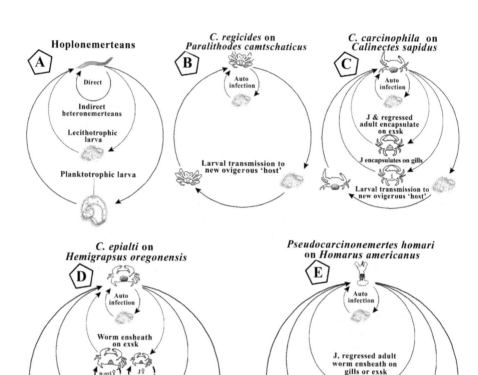

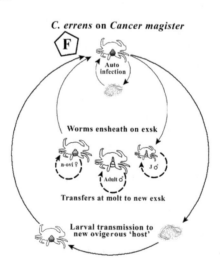

FIGURE 7.4 Contd. ...

solid gastrula changes from radial to bilateral symmetry (e.g. *Cephalothrix, Malacobdella*). (2) In heteronemerteans, the gastrula develops into a planktotrophic free-swimming ciliated pilidium (*Cerebratulus*). In them, the gastrula becomes gelatinous, transparent and more or less helmet-shaped through the extensive downward growth of oral lobes on each side of the mouth. (3) With a tendency toward lecithotrophism (Maslakova and Hiebert, 2014), the Iwata larva of *Micrura akkeshiensis* develops from a relatively flat gastrula that elongates to attain a pyriform shape. The non-feeding larva develops no bands of elongated cilia for food capture (Iwata, 1958, 1960).

The other reported variant pilidial morphotypic non-feeding larvae are: (i) Doser larva in *Lineus viridis* and (ii) Schmidt larva in *L. ruber* (von Dohren, 2015) as well as prototroch in leneid *M. rubra maculosa* (Maslakova and von Dohren, 2009) and telotroch in *M. verrilli* (Maslakova and Hiebert, 2014). For more details on structure of these larvae and their sequence of development and metamorphosis, Hyman (1951a) and von Dohren (2015) may be consulted.

Life Cycles

In most hoplonemerteans, the life cycle is simple and direct (Fig. 7.4A) but indirect in the others (e.g. heteronemerteans). In most indirect cycles, a free-swimming planktotrophic larva, the pilidium is included (Fig. 7.4A). In others, a series of non-feeding, morphologically variant, morphotypic pilidial larvae has been described.

Adaptive radiation occurs in a number of species within the genus *Carcinonemertes*, which feeds on the eggs attached to the pleopod setae of decapods. In general, the life cycle of these 'parasites' involves the shortest larval duration and longest juvenile stage. In the simplest cycle *C. regicides* larvae are transmitted to the same host by both autoinfection and other ovigerous *Paralithodes camtschaticus* crabs (Fig. 7.4B). With *Callinectes sapidus* remaining non-berried for 9 months, *C. carcinophila* juvenile migrates to the gill chamber, where it secretes a hard capsule and encysts. During the encystment, the quiescent juvenile reduces its body size and dedifferentiates its reproductive organs, and awaits the impending molt and berrying.

...FIGURE 7.4 Contd.

FIGURE 7.4

A. Direct and indirect life cycle in hoplonemerteans and heteronemerteans involving planktotrophic and lecithotrophic (for want of other figures Doser larva is shown) larva, respectively. B–F. selected life cycles of *Carcinonemertes* spp and *Pseudocarcinonemertes homari*. j = juvenile, n-ovi = non ovigerous, exsk = exoskeleton. Life cycles are shown in light continuous lines and molt cycles in dark broken lines (modified and redrawn from Kuris, 1993).

Hence, *C. carcinophila* can autoinfect as well as transmit to the new host through larval transmission (Fig. 7.4C). *Hemigrapsus oregonensis* remains berried for a couple of times with an intervening molt during summer and another couple of times but without the intervention of a molt during winter. *C. epialti* juvenile is enclosed in a soft mucous sheath along the arthropodial membrane. During the lengthy stay on the exoskeleton, it depends solely on the dissolved organic substance leaked from the crab. Following molt and berrying of the crab, it migrates to the egg mass. For more details, see Fig. 7.4D. Besides autoinfection through larval transmission on *Homarus americanus* berried for 10–12 months, *Pseudocarcinonemertes homari* juveniles and regressed adults survive ensheathed on gills or exoskeleton. Juvenile females and the non-regressing adult males survive ensheathed on gills or exoskeleton (Fig. 7.4E). Following the short-ensheathed duration, these worms migrate to the egg mass, as soon as the female lobster is berried again. The juveniles and adults *C. errens* survive ensheathed only on exoskeleton of *Carcinus magister*. However, the worms are transferred to exoskeleton, when the crab molts (Fig. 7.4F).

Endocrines and Differentiation

Thanks to Dr. Jacques Bierne, convincing evidence is provided for the existence of Gonad Inhibiting Hormone (GIH) in nemerteans. Temperate nemerteans actively breed during spring and summer but pass through a quiescent period during autumn and winter. Both the (Anopla) heteronemertean *Lineus ruber* and (Enopla) hoplonemertean *Amphiporus lactifloreus* regenerate the rest of the body following bisection and trisection. This regenerative potency provides an opportunity for experimental manipulations. The successive ingenious experiments led Bierne and Rue (1979) to recognize the brain as a precise source of GIH. Their observations listed in Table 7.1 reveal that GIH inhibits differentiation of the gonads and associated organs namely the gonoduct and sex-specific skin glands, gametogenesis and gamete maturation. Their subsequent autoradiographic studies have shown that GIH inhibits synthesis of nucleic acid and protein synthesis in gonadal cells during the quiescent period, the deprivation of GIH from the adult brain promoted RNA synthesis in young oocytes and DNA synthesis in spermatogonia, as well as induced primary spermatocytes and protein synthesis in sex-specific skin glands.

Wound healing: Nemerteans are capable of wound healing. In the clonal species *Lineus socialis* and *L. vegetes* with potency for anterior and posterior regeneration wound healing does not occur, if the fragments are less than half as long as the body width (Riser, 1974). Most authors are silent on blastema formation. However, Coe (1934) described that the healing commenced with migration of cells not only from the vicinity of the wound but also from

TABLE 7.1

Experiments that showed the brain as the precise source of Gonad Inhibiting Hormone (GIH) in nemerteans (condensed from Bierne and Rue, 1979)

Experiments	Reported Observations	
	Anoplan *Lineus ruber*	
Bisected ♀, ♂	Control	Experimental
	Anterior halves	**Posterior halves**
	Small ovaries with immature oocytes. Oviduct and ♀ glands absent	Ovaries with large oocytes. Oviducts and ♀ glands differentiated
	Small testes with no spermatogenesis. Sperm duct and ♂ glands absent.	Large testes with spermatogenesis. Sperm duct and ♂ glands differentiated
Bisected ♀ and ♂ castrated by eosophageal transection	No sexual organogenesis	Precocious sexual differentiation. Maturation of gonads with gonoducts and sex-specific glands
Trisected head with or without brain grafted to median or posterior thirds	In the anterior thirds with brainless head, precocious sexual differentiation occurs. In the median or posterior thirds grafted with brain, sexual differentiation retarded. In all those with no brain, oocytes grow rapidly, and reproductive glands and skin differentiated.	

Enoplan *Amphiporus lactifloreus* : oocyte number/µm diameter					
Control		Decapitated	Trisected		
Range	Mean		Anterior	Median	Posterior
0–40	27	-	11	43	10
80–120	2	4	8	-	14
200–240	-	3	-	-	-

"dormant cells situated in the parenchyma" of "all parts of the body". It is not known whether these dormant cells are analogous to annelidan neoblasts.

Regeneration and Cloning

Regeneration: Most investigated nemertean species have the potency to regenerate, especially the regions behind the foregut. Notably, the fragment anterior to the brain anlage is not regenerated. Clearly, a part of the brain or nerve cord is required to induce regeneration. Further, the tendency for posterior regeneration has demonstrated the existence of a species-specific axial gradient, which is expressed in different degrees. The posterior fragment containing the rhynchocoel, proboscis, intestine and lateral nerve cord, completes anterior regeneration in *Prostoma graecense*, if the worm is amputated behind the brain (Riser, 1974). The regenerative potency of *L. ruber* and *L. sanguineus* was so extensive that it was possible for Bierne (1990) to generate various kinds of chimeras by grafting body parts from different individuals and species, as well. The studies on posterior regeneration in immature *L. lacteus* and 'ready to spawn' *L. ruber* suggest the existence of undifferentiated cells. In fact, a 5-month old headless regenerating *L. marinus* is reported to shed gametes. The fact that PGCs arise from the mesenchyme and the preformed gonoduct is not derived from the gonads indicates that the PGCs and other undifferentiated cells for organogenesis can be derived from the existing mesenchyme cells of the posterior fragment, especially in the absence of anterior fragment including the brain. In fact, the brain is the source of GIH to inhibit differentiation of gonad and associated sex organs, gametogenesis and maturation of gametes (Bierne and Rue, 1979).

Clonal multiplication: Of 1,300 nemertean species, clonal multiplication is described in four species, of which *Lineus pseudolacteus* and *L. sanguineus* have been confirmed as cloners (Ament-Verlasquez et al., 2016). Interestingly, these leneids are capable of both anterior and posterior regeneration (Riser, 1974, Fig. 7.5). The high levels of clonal potency and extreme rarity (but not total absence) of mature females, both in the field and laboratory, have led to suggest that a small fraction of *L. sanguineus* population may undergo sexual reproduction, which may be adequate to erase all the negative effects of clonal multiplication (Hartfield, 2016). In *L. socialis*, the fragments tend to undergo encystment, in which clonal multiplication is completed (Riser, 1974). With regard to *L. pseudolacteus*, molecular studies have revealed that with an extremely high level of polymorphism, *L. pseudolacteus* may be a hybrid triploid, arising from fertilization between a diploid *L. sanguineus* egg and haploid *L. pseudolacteus* sperm. In them, allopolyploidy may have hampered meiosis and thereby enforced *L. pseudolacteus* to undertake clonal

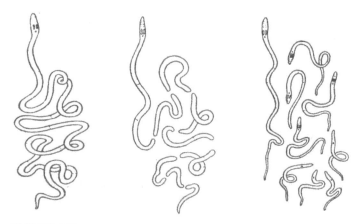

FIGURE 7.5

Sequence of clonal multiplication in *Lineus socialis*. Note the retention of polarity and axial gradient in the resulting clones (free hand drawings from Hyman, 1951a).

multiplication alone (Ament-Verlasquez et al., 2016). In fact, hybridity, polyploidy and clonality are not uncommon in nature (Lovell et al., 2013).

Starvation: In this context, it is necessary to bring the following to the attention of modern zoologists. An amazing anecdote narrated by Hyman (1951a) is that starvation and its effect on rejuvenization of nemerteans. Some nemerteans survive prolonged starvation lasting for months and years (Coe, 1943). The starvation leads to size reduction, dedifferentiation and toward reguvenization, recalling similar feature in turbellarians (see Pandian, 2020). A large number of mesenchymal cells dedifferentiate into wandering phagocytes that devour cells of the body, gut and gonad. On disintegration, the phagocyte loaded with nutrients provides nutrients for the remaining cells in a starving worm. In this manner, much of body cells disappear, while the nerve cells persist for some more time to come (Dawydoff, 1924). In *Lineus*, the starved worm is reduced to a size of *Paramecium* and resembles a pilidium larva. With starvation lasting for more than 2 years, the worm is reduced to an ovoid morula with an outer and inner epithelial layers enclosing loose mass of large round cells. It is not known whether the morula can be revived by providing the required liquid food. If the morula can osmotrophically absorb nutrients and revive, it may provide an excellent opportunity to study the sequence of regeneration and rejuvenization. It will also be interesting to bisect the morula to know whether the individual or cluster of cells is pluripotent.

8

Gnathostomulida

Introduction

The gnathostomulids are slender, cylindrical (40–100 µm in diameter), microscopically small (e.g. length: from 230 µm in *Problognathia minima* to 3,600 µm in *Haplognathia beliziensis*, Schmidt-Rhaesa, 2014), acoelomate worms. Their body is covered with a single layered epidermis, in which cells are characterized by one cilium per cell. Some epidermal cells may also contain mucous and adhesive glands. In them, the subterminal mouth leads to a to a muscular pharynx armed with paired pincer-like jaws and an unpaired jugum, and a straight gut with no anus, albeit some may possess a 'temporary anus' (Hejnol, 2015c). The nervous system, located intraepithelially, consists of frontal and buccal ganglia and longitudinal interconnected nerve fibers. There is no body cavity and mesenchyme (Schmidt-Rhaesa, 2014).

Taxonomy and Distribution

In 1974, Sterrer reported the existence of 80 species in 18 genera and predicted that the total number would exceed 1,000 species. However, only 100 species have so far been described (Schmidt-Rhaesa, 2014, Hejnol, 2015c), i.e. the number of newly erected species has increased at the rate of 0.5 species/y. With hermaphroditism and direct life cycle, and a consequent lifetime fecundity limited to 1–6 eggs, it is not likely that the gnathostomulid species number increases further. Gnathostomulids are classified into two orders: Filospermoidea with filiform sperm (e.g. *Haplognathia*, Fig. 8.1A) and Bursovaginoidea. The latter is further divided into two suborders: Scleroperalia with cuticularized bursa (e.g. *Gnathostomula*, Fig. 8.1B) and Conophorlia with conulus-type sperm (e.g. *Austrognatharia*, Fig. 8.1C) (Sterrer, 1974). For more details on classification of gnathostomulids, Jenner (2014) may be consulted.

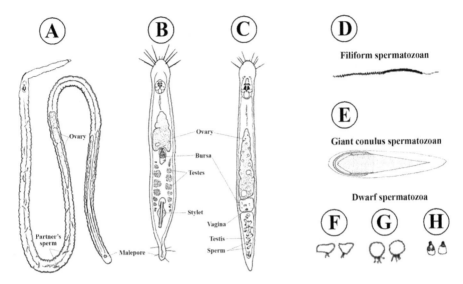

FIGURE 8.1

Dorsal view of A. filospermoidean *Haplognathia simplex*, B. scleroperalian *Gnathostomula jenneri*, C. conophorlian *Austrognatharia kirsteueri*. Spermatozoan of D. filospermoidean filisperm of *Haplognathia simplex*, E. conophorlian giant sperm of *Austrognatharia riedii* and dwarf spermatozoa of F. *Agnathiella beckeri*, G. *G. jenneri* and H. *Onychognathia filifera* (free hand drawings from Sterrer, 1974).

Gnathostomulids are exclusively marine inhabitants. Some of them feed on decaying plant leaves; others are interstitial grazing on chemoautotrophic sulfur-oxidizing bacteria. The latter occur up to the depth of 400 m and are common just below the oxidized layers of thiobios in the sandy bottom (Schmidt-Rhaesa, 2014).

Reproductive Biology

Gnathostomulids are simultaneous hermaphrodites. The female system consists of an unpaired elongated pear-shaped ovary located dorsally between the gut and epidermis. The ovary holds 4–6 oocytes, whose size increases posteriorly with the last mature oocyte considerably larger (50–60 μm) than the others. In bursovaginoideans, the mature ooctye number is just one. The ovary is continued as bursa system, which is simple in *Conophoralia* but complicated in Scleroperalia. The male system consists of a paired (e.g. Scleroperalia) or unpaired (e.g. *Conophoralia*) testes, vasa deferentia, a penis and a gonopore. Reciprocal insemination may occur, especially in Filospermoidea. Sterrer (1974) recognized three types of copulations: (i) sperm penetration into the partner's skin at the mid-dorsal region, where the vagina is located, (ii) the hypodermic injection and sperm storage in the bursa, and (iii) injection into the vagina. Seasonal body fragmentation is the suspected route, through which the fertilized eggs are released (e.g. *Gnathostomalia lutheri*). In *G. jenneri*, cleavage of the spherical zygote commences with the first and second divisions meridional (from pole to pole), nearly equal and holoblastic. From the third division onwards, it occurs in alternation of dextropic and leiotropic planes. For more details on cleavage and embryonic development, Sterrer (1974) and Schmidt-Rhaesa (2014) may be consulted. In gnathostomulids, development is direct.

Part C
Pseudocoelomata

The pseudocoelomates include minor phyla only. Rotifera (2,022 species) and Nematoda (27,333 species) are speciose phyla but others are not. Members of all the six phyla are eutelic, i.e. after hatching, mitosis is ceased in their somatic cells, a feature that has not been hitherto adequately recognized. In them, the eutely in them has eliminated the scope for regeneration and clonal reproduction. The pseudocolomates are gonochores and reproduce sexually. Sexual reproduction in them is supplemented by parthenogenesis. Surprisingly, some of these eutelics survive extremes of temperatures (up to 61.3°C, desiccation, frozen) and in the absence of oxygen. Their life cycle is direct except in Nematomorpha and Acanthocephala. All the members of Nematomorpha (360 species) and Acanthocephala (1,100 species) and 59% of Nematoda are endoparasites. Their life cycle may involve one or more intermediate hosts. The first two comprise parasites alone. Parasitism has led to degenerating only some pharyngeal structure(s) in nematodes.

9

Rotifera

Introduction

The 2,022 speciose rotifers are saccate or cylindrical, tiny (50 µm to 2,000 µm in length) eutelic pseudocoelomates. Their fluid-filled body cavity is not bounded by an epithelial layer and is thus not a true coelom (Hejnol, 2015a). Their body is composed of a head, neck (only in some species), a trunk and foot. Although folds may occur in the body, they are not segments; thus rotifers are not metameric. Their body wall (integument) contains a filamentous layer of varying thickness called the Intra Cytoplasmic Layer (ICL). In many species, the integument is strengthened by a thick ICL called lorica. As 'wheel bearers', the rotifers are characterized by an anterior corona composed of an outer troch and an inner singulum of concentric rows of cilia. The corona leads to the mouth, muscular pharynx bearing a mastak with a complex set of jaw apparatus called the 'trophi', stomach and intestine, which opens through the anus. Rotifers are mostly microphagous suspension feeders or grazers; some are raptorial predators. All of them use the corona for food collection but free-swimming species also engage it for locomotion. In non-feeding dwarf males, the rudimentary gut serves as an energy source for fast swimming (Wallace and Snell, 2010). Rotifers have an anterior brain and a pair of lateral nerves and excretory protonephridia. The important cytological characteristic of rotifers is the eutelic constancy of the number of cells after hatching (Stelzer, 2005), and presence of syncytial cells and organs (Gilbert, 1993). Some rotifers act as vectors of virus that attack shrimp. In aquaculture, rotifers play an important role as live feed next only to *Artemia* nauplii (Lubzens et al., 1989). They are highly nutritious. In China, Israel and Japan, aquaculturists have utilized rotifers maximally in intense culture systems of planktivorous fish.

Taxonomy and Life Style

On the basis of a paired or median gonad(s) and other characteristics, the phylum Rotifera is classified into two classes: Pararotatoria and Eurotatoria; the latter is divided into two subclasses: Monogononta and Bdelloidea. Each of these subclasses is further divided into three orders (Fig. 9.1). The Pararotatoria compose a single family Seisonidae with dispecific *Seison* and monospecific *Paraseison*. The most speciose (1,570 species) Monogononta include 128 genera in 28 families. With 461 species in 20 genera, Bdelloidea is classified into three families (Segers, 2008).

Rotifers are predominantly limnetic inhabitants, albeit a few are limnoterrestrials, inhabiting on mosses, lichens and liverworts in moist soil (e.g. Glime, 2017a). According to Segers (2008), 74 of 2022 species, i.e. 3.65% are marine rotifers (however, see Fontaneto et al., 2008). Their distribution in eight geographic regions is listed in Table 9.1. The majority of rotiferan genera and species occur in the Palearctic region. Among rotifers, 30 genera are monospecific; ~ 235 genera are more speciose. The notommatid *Cephalodella* (159 species) and *Lacane* (200 species) are the most speciose genera. Remarkably, 900 species are endemic but 405 are cosmopolitan. The marine rotifers include only two genera *Seison* and *Paraseison*, and the bdelloid

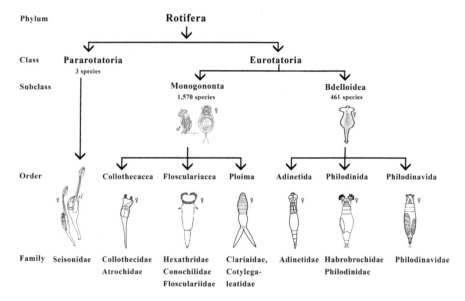

FIGURE 9.1

Classification of Roifera. Note the size of seisonid and monogonontan males. And also note the pseudosegmentation in bdelloids (condensed from Wallace and Snell, 2010, all figures are free hand drawings from different sources).

TABLE 9.1

Distribution of rotifer species and genera (condensed from Segers, 2008)

Geographic Region	Genera (no./family)	Specie (no./family)	Marine Species
Palearctic	113	1350	70*
Afrotropical	85	917	1**
Australian	96	682	3***
Oriental	87	591	
Nearctic	109	544	
Neotropical	91	687	
Antarctic	29	133	
Pacific	47	91	
Total/Mean	657/82	1948/244	74

Marine: * = Monogononta, ** = Bdelloidea, *** = Seisonidae

Zelinkiella (perhaps *Anomopus* also, Wallace and Snell, 2010). All other genera include freshwater inhabitants, but their representatives are rooted from the sea. The genera *Encentrum* and *Synchaeta* have more species in brackish and sea waters than they do in freshwater (Thane, 1974). Rotifers usually occur in abundance; their density ranges from $32,000/m^2$ in limnoterrestrial species to > 2 million/m^2 in aquatic habitats (Wallace and Snell, 2010). A vast majority of rotifers are solitary. However, ~ 25 species belong to eight genera in Monogononta are 'pseudocolonials' (Wallace and Snell, 2010). All these quantifications are based on morphotaxonomy. Species boundaries of rotifers are based on chemical communication. Many reproductively isolated taxa are simply lumped into a single species (Snell et al., 2006). Molecular techniques have begun to reveal the existence of several genetically divergent lineages, for example, in the *Brachionus plicatilis* complex (see Wallace and Snell, 2010). It is likely that the species number may increase, especially in monogononts. However, the morphotaxonomy is still valid, as a cross between *Brachionus urceolaris* female and *B. quadridentatus* male produced not a hybrid but *B. urceolaris* female progenies alone. In them, the chromosomes of *B. quadridentatus* were eliminated and only *B. urceolaris* amictic females were produced through cryptoparthenogenesis (Birky and Gilbert, 1971).

All the three seisonids are sessile and epizoic only on the leptostracan crustacean *Nebaliae*. *S. annulatus* occurs on the pleopods, while *S. nebalea* on the pereiopods (Ricci et al., 1993). Most rotifers can freely swim (0.18 to 1.04 mm/second [s], Starkweather, 1987, Rico-Martinez and Walsh, 2013); some of them can even jump. However, sessile rotifers are found in three families (Fig. 9.1): Atrochidae (2 genera), Collothecidae (5 genera) and Floscularidae (7 genera), i.e. 3.2% of rotifers are sessile. Their densities may reach as high as > 6 individual (indivi)/mm^2 (Wallace and Snell, 2010). With a parthenogenic mode of reproduction alone, almost all the bdelloids are free swimmers to disperse their clonal offspring. Most rotifers are oviparous and release their eggs into pelagic waters (e.g. *Notholca*); some planktonic species either carry their eggs attached by a thin thread (e.g. *Brachionus*) or fix them to substratum (e.g. *Euchlanis*). The sessiles carry their eggs on a specialized structure called oviferan (e.g. *Sinantherina*). Limnoterrestrial rotifers are easily infected by endoparasitic fungi. Planktonic rotifers are infected by sporozoans. Rotifers can ingest the oocytes of parasitic protists of importance to human health. A few rotifers are parasites. For example, *Proales gigantea* is a parasite of the eggs of pond snails. Nearly all members of the notommatoid genera *Dilophaga*, *Albertia* and *Balatro* are obligatory parasites on freshwater oligochaetes and leeches. In them, parasitism reduces the head, corona, sense organs and mastax as well as alternation of the foot into an organ of adhesion (Hyman, 1951b). Numerous rotifer species are reported from hot springs; they can survive temperatures up to 46°C (e.g. *Philodina roseola*, see Hyman, 1951b).

Eutelism and Tissue Types

A notable characteristic of rotifers is that (i) all their organs are syncytial and (ii) their number of cells or rather nuclei is constant. For example, the common *Epiphanes senta* has the following nuclei number:

Ectodermal derivities	No.	Endodermal derivities	No.	Mesodermal derivities	No.
1. Coronal epidermis	172	8. Esophagus	15	12. Circular muscles	22
2. Trunk & foot epidermis	108	9. Stomach	39	13. Retractor muscles	40
3. Mastax epidermis	91	10. Gastric glands	6+	14. Mastax musculature	42
4. Brain	183	11. Intestine	14	15. Protonephridium	28
5. Peripheral nervous system	63			16. Ovary	1+
6. Mastax nerve cells	34			17. Vitellarium	8
7. Pedal glands	19			18. Oviduct	3
Subtotal	670		74+		144+

In almost all rotifers, the number of nuclei is fixed as six for gastric glands and eight for vitellarium. In all, the total number of nuclei ranges between 900 in *E. senta* to 959 in *Asplanchna priodonta* for monogononts. It is, however, 428 only for bdelloids (e.g. *Habrotrocha rosa*, Hyman, 1951b). Remarkably, the number of tissue types from ectodermal derivatives 7+, endodermal derivatives 4+ and mesodermal derivatives 7+ may not exceed 20. It is a few more than that (14 tissue types) in turbellarians (Pandian, 2020).

A consequence of eutelism is that "the female reproductive system consists of a single syncytial ovary and a syncytial vitellarium found together in a common membrane that continues to a cloaca as simple tubular oviduct" (Hyman, 1951b, Fig. 9.2). The vitellaria and oviducts are paried in Bdelloidea. In contrast, the seisonacids have paired ovaries without vitellaria. The male reproductive system consists of a single large sacciform testis, from which a ciliated sperm duct proceeds to the gonadal pore terminating in a protrusible penis. The males are progressively reduced to a 'sacculinid'-like body (Fig. 9.2). A second consequence of eutelism is that the oocyte number is determined as 32 in *A. priodonta*, 14 in *A. girodi* and 8–10 *Testudinella patina* (Gilbert, 1993).

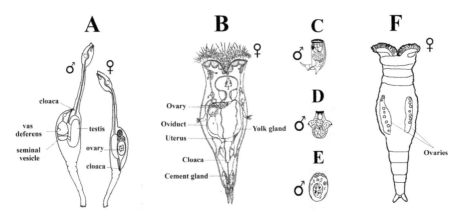

FIGURE 9.2

Reproductive system in A. gonochoric pararotatorian *Seison nebaliae* (redrawn from Ricci et al., 1993), B. monogonontan female and males showing progressive reduction in C. *Mytilina mucronata*, D. *Pedalia mira* and E. *Asplanchna sieboldi*. F. bdelloid female (free hand drawings from different sources).

Reproduction

The seisonids are gonochores; their female and male ratio remains approximately equal. Notably, their males are larger than females (Fig. 9.2A). In them, the mating involves the assured transfer of spermatophores by the male on the body surface of the neighboring female (Gilbert, 1993). With no males, the bdelloids are resorted to parthenogenesis alone. In monogononts, three types of females are recognized: (i) Amictic (or ameiotic) stem female that produces clones solely through parthenogenesis; clonal female, in its turn, becomes an amictic stem female. (ii) When conditions are favorable and at high densities, amictic females become mictic, undergo meiosis and produce haploid sexual mictic eggs. Mictic females can be impregnated with sperm only during the first few hours after hatching/birth and prior to the cuticle hardening (Birky and Gilbert, 1971). Resting eggs are generated from fertilized mictic eggs but males from unfertilized mictic eggs. In contrast to seisonids, the monogonont male is smaller than the female; the non-feeding males lack the gut and other organs (Fig. 9.2C, D, E) to such an extend that the *Asplanchna sieboldi* male is only a sac filled with ~ 50 sperms. In seisonids, the male has paired saccular testes, which terminate into cloaca through seminal vesicle and vas deferens (Ricci et al., 1993). (iii) In the rotifer female, the ovary is a small, syncytial mass, which is closely associated with vitellarium (yolk glands). Both the ovary and vitellarium are enclosed by a follicular layer, which continues as oviduct and terminates into a cloaca. In bdelloids, the paired ovaries are located bilaterally but a median single ovary is present in monogononts.

Life Cycle

Rotifers are characterized by a simple and direct life cycle. In seisonids (Fig. 9.3A) and bdelloids (Fig. 9.3E), neonates are hatched from fertilized and parthenogenic diploid eggs, respectively. In monogononts also, the neonates are hatched from amictic or mictic eggs. However, their life cycle is complex with heterogony, i.e. the cyclic parthenogenesis is alternated with sexual reproduction (Fig. 9.3B). Incidentally, cyclical parthenogenesis is a mode of reproduction that combines the advantages of rapid clonal propagation through diploid ameiotic parthenogenesis and sexual mictic reproduction (Serra and King, 1999). The parthenogenic cycle may be monocyclic, as in ephemeral rock pool (hueco)-inhabiting *Hexarthra* sp (Schroder et al., 2007) or dicyclic, the first in spring and the second in autumn, as in most temperate species or 12 generations or more (e.g. *Brachionus*, Gilbert, 2002). As a result, as many as 10^{12} amictic females can be generated within 60 days (Ruttner-Kolisko, 1963) to rapidly colonize a favorable habitat. The heterogonic life cycle can be further complicated with the production of physiologically and morphologically distinct subitaneous and diapausing eggs within the parthenogenic cycle (Fig. 9.3C). In *Synchaeta pectinata*, Gilbert (1995) reported the production of a thin (~ 1.4 µm) shelled subitaneous eggs that hatch without the arrest of diapause and thick (9 mm) shelled eggs that enter into obligate diapause between the first and third cleavages. Diapause is induced by brief starvation and represents an adaptation to tide over short durations of low or no food availability. It lasts for 14 days at L16: D8 photoperiod and 19°C. But it is not broken below 5°C. Another complication reported is the presence of amphoteric females, capable of alternatively producing amictic and mictic eggs (Fig. 9.3D). It must, however, be noted that the number

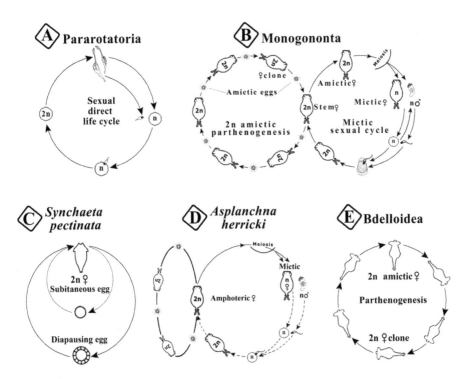

FIGURE 9.3

Rotiferan life cycles: A. sexual direct life cycle in *Seison nebaliae* (based on Ricci et al., 1993), B. typical heterogonic life cycle in Monogononta, C. subitaneous and diapausing egg production within the 2n mictic parthenogenic cycle of *Synchaeta pectinata* (modified and redrawn from Gilbert, 1995), D. life cycle of *Asplanchna herricki*, in which the same amphoteric female lays both amictic and mictic eggs. As information on gamete production and fertilization is not yet available, a tentative one is drawn with dotted lines (based on Gilbert, 1974, Rico-Martinez and Walsh, 2013) and E. typical parthenogenic life cycle in Bdelloidea.

of amphoteric species and species capable of producing subitaneous and diapausing amictic eggs are few and more an exception.

The amictic and mictic monogonont females produce three types of eggs. In *Brachionus patulus*, for example, the size is the largest (123 µm in diameter) in fertilized mictic egg, intermediate (106 µm) in amictic egg and the smallest (66 µm) in unfertilized mictic egg (Rao and Sarma, 1985). Figure 9.4 shows the amictic, unfertilized and fertilized mictic eggs of *Epiphanes*. The embryonic development of rotifers is distinctly different from that of other low invertebrates. The early cleavages suggest the determinate spiral type and are comparable to that in Acanthocephala. The first cleavage commences with an unequal division, giving rise to a large blastomere called the CD in the vegetal pole and smaller AB blastomere in the animal pole (Fig. 9.4IV–X). The descendants of CD give rise to the germovitellarium and other internal organs. The second division cleaves equally between A

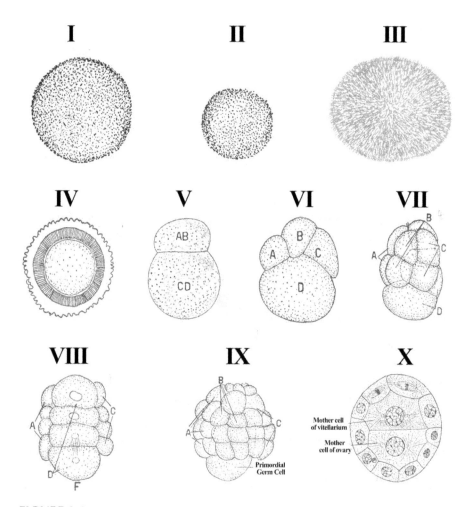

FIGURE 9.4

Amictic (I), unfertilized (II), and fertilized (III) mictic eggs of *Epiphanes*. Early embryonic development (IV–X) in a resting egg (free hand drawings from Hyman, 1951b).

and B blastomeres but unequally C and D blastomeres, of which D is the largest and cleaves more rapidly. The third division leads to a 10-cell embryo. Interestingly, the D blastomere is the Primordial Germ Cell (PGC). Hence, the germinative and somatic lineages are determined early at the 4-cell stage itself. An experimental study has indicated that the fate of blastomeres is determined early and that the blastomeres differentiate into the final fates without induction by neighboring cells. Therefore, the rotifer embryogenesis is a classic mosaic type. For more details on organogenesis, Hejnol (2015d) may be consulted.

Fecundity and Mixis

Among different live feed types, alga (*Chlorella vulgaris*) fed either at low or high density sustain rapid growth and increase in population density up to the age of 12 days in *Brachionus calyciflorus* (Fig. 9.5). In general, the life span of rotifers is short and ranges from 3–8 days in *Proales decipions* to 40 days in *Cupelopagis vorax* among monogononts and 10 days in *Philodina roseola* to 5 months in calidinid bdelloids. Their fecundity ranges from 14 eggs in *Lecane inermis* to 40 neonates in *C. vorax* among monogononts and from 2 eggs in *Adineta barbata* to 6 eggs in *Haprotrocha constricta* among bdelloids (Hyman, 1951b). Bdelloids have a longer life span but are less fecund than monogononts. Prior to the description of fecundity, a preamble is required. Considering the basic life history traits like egg size, generation time, size at maturity and so on, Stelzer (2005) arrived at the following important conclusions: (1) Rotifers are eutelic, i.e. there is no cell division after hatching. Hence their growth is solely based on cytoplasmic expansion. However, with a difference in the initial cell number at hatching, the body volume of *Asplanchna priodonta* with larger intercellular spaces is as much as 2,000-fold larger than the densely packed *Keratellacochlearis tecta*; this difference is reduced to 87-fold in terms of carbon content (Telesh et al., 1998). Yet, the fixed cell number at hatching sets upper limits to the body size. (2) More than 96% rotifers are motile by ciliary propulsion, which may impose a constraint on maximum body size, as it becomes energetically inefficient in larger body sizes. (3) Clutch size (Batch Fecundity, BF) is fixed in all rotifers at one egg/batch. With a pair of ovaries, the bdelloids also release their eggs sequentially, alternatively between the two ovaries. Even

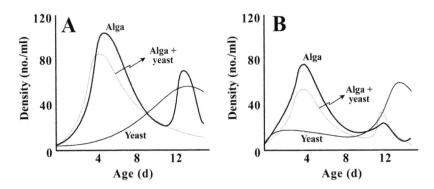

FIGURE 9.5

Effect of food types namely alga (*Chlorella vulgaris*) and/or yeast (*Saccharomyces cerevisiae*) on population growth of *Brachionus calyciflorus* fed on A. high food density of 3×10^6 cells/ml and B. low food density 1×10^6 cells/ml (compiled from Sarma et al., 2001).

in the so called pseudoclutch of *Brachionus*, the sequentially produced eggs of different ages are brooded.

Among life history traits, trade off can occur; with an increase in a trait like current reproduction may cause a decrease in the future survival and/or fecundity representing a cost of reproduction (e.g. Sarma et al., 2002). The reproductive cost is specific to egg bearing species, in which the swimming performance is directly decreased with increasing number of eggs carried. Consequently, it results in early reproduction in a smaller size with low fecundity or late reproduction at a larger size with greater fecundity. Understandably, the first alternative is chosen by rotifers that inhabit unpredictable habitats and the second by rotifers living in stable and favorable habitats. From a pioneering study with 21 different clones of *Asplanchna brightwelli*, Snell and King (1997) derived two basic patterns: (i) some clones produced a lot of eggs but lived for a short period, while the others reproduced slowly but lived longer. However, it must be noted that the environment may place a rider over these alternatives; for example, *Hexarthra* sp inhabiting ephemeral rock pools may produce fewer eggs but live only for a short period.

With reference to reproduction, it is customary to divide the life span into (i) pre-reproductive, i.e. Generation Time (GT), (ii) reproductive and (iii) post-reproductive (menopause) phases. Most authors include the embryonic duration also into the GT of rotifers (e.g. Schroder et al., 2007). While a number of *Asplanchna* species include no menopause period, *Proales sordida* spends 14.6 of 184 hours or 8% of its Life Span (LS) on senile post-reproductive period (see Birky and Gilbert, 1971). Temperature alters GT considerably. For example, GT as a percentage of LS is 28, 48 and 87 in tropical, temperate and arctic crustaceans, respectively (Pandian, 2016). Temperature profoundly alters embryonic duration (Fig. 9.6A) and GT (Fig. 9.6B). Expectedly, survival of rotifers progressively decreases with

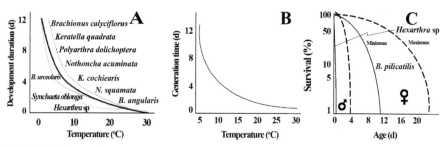

FIGURE 9.6

A. Embryonic development duration and B. generation time as a function of temperature in some monogononts. The trend is drawn for the species named in Fig. 9.6A (modified and redrawn from Schroder et al., 2007). C. Survival of *Brachionus plicatilis* as a function of age at 25°C and *Hexarthra* sp (compiled and redrawn from King and Miracle, 1980, Schroder et al., 2007).

advancing age (e.g. *Brachionus plicatilis*). Depending on the environmental conditions, there can be minimal and maximal survival trends. In *B. plicatilis* females, the minimum and maximum LS are 12 and 22 days, respectively. But it is just 2 and 3 days for males (Fig. 9.6C).

In rotifers, fecundity is measured by the number of neonates or offspring rather than the number of oviposited eggs. Hence, the reported values for the neonates may be more than the actual number of eggs laid. It may be appropriate to introduce some terms related to fecundity, as used by fishery biologists (Pandian, 2010). Batch Fecundity (BF) is the number of eggs laid per spawning. Most reported values for the rotifers are the Lifetime Fecundity (LF), i.e. the number of neonates produced by a female during her entire lifetime. For a rotifer species, the Potential Fecundity (PF) is fixed at birth. The approximate PF has been estimated as 32 oocytes for *Asplanchna priodonta*, ~ 20 for *Synchaeta grimpei*, *S. baltica* and *S. travina*, 14 for *A. girodi* and 8–10 for *Testudinella patina* (see Gilbert, 1993). However, due to limiting factors like food availability, the Realized Fecundity (RF) is always less than that of PF. In rotifers, most investigations have reported the role played by body size, algal size and/or density and temperature in realizing the PF. BF is a function of body size (L). Because BF is related to the volume of space available in the pseudocoelomate cavity to accommodate the ripe ovary and eggs ($F = {_a}L^b$), geometry suggests that the length exponent would be 3.0. However, as rotifers lay a single egg at a time, the length exponent may not be relevant. Nevertheless, it is relevant to LF. Considering values for 35 planktonic rotifer species, Walz et al. (1995) arrived at a linear positive relation between egg volume and body volume (Fig. 9.7A). Wallace et al. (1998) found a similar relation for the sessile collothecids and flosculariids. Three points could be noted: (i) the planktonic species have a wider size range than the sessiles. (ii) The level of body size-fecundity relation is at the highest for the freely drifting planktonic species, indicating that they are relatively more fecund than sessile rotifers. The flosculariids may have relatively a longer juvenile swimming duration than collothecids, which are able to save and allocate more resources for reproduction. (iii) More importantly, the ratio of egg volume to body volume (REV = Relative Egg Volume) does not stay constant, despite the linear and positive relation between body size and fecundity. Instead, small rotifer species consistently show a larger REV than larger species (Walz et al., 1995). This basic finding has also been confirmed independently in collothecid and flosculariid species (Wallace et al., 1998) and in several marine *Synchaeta* species (Rougier et al., 2000). In bdelloids too, this has been confirmed by Ricci and Fascio (1995, however, see Ricci, 1995). Hence, small rotifer species make a higher investment on individual offspring than larger rotifers. The former may also grow less to attain sexual maturity, i.e. the smaller rotifers have shorter GT. This finding has a lot of implications to rotifer culture.

In rotifers like *B. plicatilis*, offspring production or LF grows with increasing algal size but decreases with algal density (Fig. 9.7D). The egg

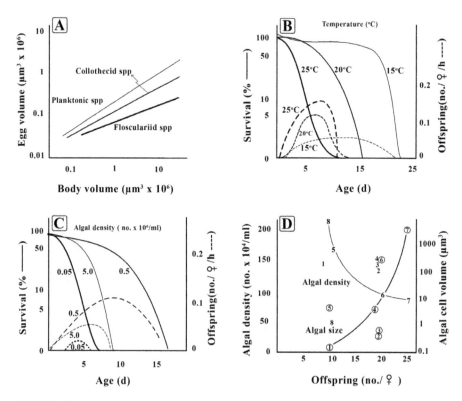

FIGURE 9.7

A. Egg size as a function of body size in 35 planktonic species, flosculariid 29 sessile + 6 planktonic species and collothecid 24 sessile + 6 planktonic species (compiled and redrawn from Walz et al., 1995, Wallace et al., 1998). B. Effect of temperature on survival and offspring production in *Brachionus calyciflorus* (compiled from Halbach, 1970). C. Effect of *Chlorella pyrenoidosa* density (no. × 10⁶/ml) on survival and fecundity of *B. calyciflorus* (Halbach and Halbach, 1973). D. Effect of algal density and algal size on lifetime fecundity of *B. plicatilis*: 1. *Nitzchia clostrium*, 2. *Monochrysis lutheri*, 3. *Dunaliella tertiolecta*, 4. *Chlamydomonas* sp, 5. *Cyclotella cryptica*, 6. *Chlorella* sp, 7. *Synechococcus elongatus*, 8. *Eutreptiella* sp; values for algal size are shown by circled numbers (drawn from data reported by Hirayama et al., 1979).

production rate is inherently associated with survival, as has already been indicated earlier. In *B. calyciflorus*, for example, 50% of the cohort survives to age 16 days at 15°C, 11 days at 20°C and 5 days at 25°C. Further, fecundity is compressed into fewer age classes at higher temperatures. The daily offspring production totals to 13 neonates at 15°C, 16.6 at 20°C and 12.9 at 25°C (Fig. 9.7B). Correspondingly, the offspring production rate is extended from the age of ~ 8 days to 22 days, albeit the rate is low at ~ 0.05 neonate/female/d. At 25°C, the rate is high 8–10 neonate/female/d but lasts only up to the age of 11 days. Similarly, the LS lasts for 18 days, when fed at 0.5 *Chlorella pyrenoidosa* cells × 10⁶/ml and the offspring production rate is spread over a longer duration than those fed at either low density of 0.05×10^6/ml or super optimal density of 5.0×10^6/ml (Fig. 9.7C). The values for bdelloids are 0.1, 1.1, 1.31 (but 1.22 egg/female/d for *Philodina roseola*, Moreira et al., 2016) and 2.74 egg/female/d for *Rotifer vulgaris*, *P. citrina*, *P. roseola* and *P. acuticornis*, respectively (see Gilbert, 1993). Briefly, body size, temperature, algal size and density have profound effects on fecundity of amicitic females.

Some of these experiments have provided adequate evidence for the production of eggs at different ages. For example, the 3-day old *Synchaeta* females produce larger eggs than 2-day old females (Kirk, 1997). It may be beneficial to produce larger starvation-resistant neonates at low food densities (Glazier, 1992). At low temperatures, larger eggs are produced than at high temperatures. Reared at 4°C, *S. pectinata* produced 35% larger eggs, in comparison to those produced at 12°C. But the neonates from large eggs did not exhibit a greater fitness than those arising from small eggs (Stelzer, 2002). Hence, it may be interesting to see whether small and large eggs are also resistant to desiccation. In bdelloid *Macrotrachela quadricornifera*, differently sized eggs were produced by feeding the mother rotifers at suitable algal densities. The ability to survive over stress like desiccation was not affected by egg size. However, the neonates hatched from large eggs ($242 \ \mu m^3 \times 10^3$) with a shorter duration of GT (4.7 days) are not favored under harsh conditions. The neonates hatched from a small egg ($177 \ \mu m^3 \times 10^3$) pass through longer embryonic duration (5.2 days) and GT (5.5 days). The recovery success after desiccation is 93 and 95% for neonates appearing from small and large eggs, respectively (Santo et al., 2001).

In the pseudocoel-developing embryos of *Brachionus* species, mixis is determined prior to the commencement of cleavage, or event at the oocyte maturation stage (Hyman, 1951b). However, the determination remains labile for long during embryonic development *in utero* developing *Asplanchna* species (Birky and Gilbert, 1971). As a sole food source, *Paramesium caudatus* fails to induce mixis in *Aplanchna*. For example, the *C. vulgaris*-fed *B. rubens* produce 13.6, 19.4 and 7.2 amictics, unfertilized mictics and fertilized mictics, respectively (Table 9.2). These values may be compared with 6.4, 12.8 and 5.7 amictics, unfertilized mictics and fertilized mictics of *B. rubens* fed with *Aerobacter aerogenes*. In *B. rubens*, the mean interoviposition period is also extended from 0.45 days in *C. vulgaris* fed amictics to 0.62 days in *A. aerogenes*-fed amictics. However, the differences are narrowed in mictic females (Table 9.2).

The switch from amictics to mictics is keyed by population density, photoperiod or α tocopherol, the vitamin E. A long day length of > 15 h/d or a photoperiod or 15–24 hours triggers mictic female production in *Notommata copeus*, *N. codonella* and *Trichocera rattus* (see Birky and Gilbert, 1971). There are two patterns of mixis induction: (i) Some rotifers like *Asplanchna girodi* follow the bang-bang pattern, in which population reaches the maximum size by producing only amictic females and then switches to an exclusive production of mictic females (Serra and Carmona, 1993).

TABLE 9.2

Effect of food type and female type on fecundity and inter-oviposition interval of *Brachionus rubens* fed on *Chlorella vulgaris* or *Aerobacter aerogenes* at 21°C (modified from Pilarska, 1972)

Species	Fecundity (no./♀)	Interval (d)
Amictic/*C. vulgaris*	13.6	0.45
Unfertilized mictic/*C. vulgaris*	19.4	0.33
Fertilized mictic/*C. vulgaris*	7.2	0.68
Amictic/*A. aerogenes*	6.4	0.68
Unfertilized mictic/*A. aerogenes*	12.8	0.34
Fertilized mictic/*A. aerogenes*	5.7	0.72

Hence, there is no overlapping between outgoing and incoming populations (Wallace and Snell, 2010). However, most rotifers follow the intermediate pattern, in which the population consists of both mictic and amictic females. With regard to rotifer density triggering mixis, it is established that at penultimate high densities, mictic females begin to appear. In *Brachionus urceolaris*, populations characterized by intermediate pattern (Serra and King, 1999), for example, include (i) amictic females with LS of 9 days and fecundity of 20 females, (ii) 25 amictic males with 9.5 days LS and (iii) fertilized mictics with 10 days LS and fecundity of 4 resting eggs. Hence, the proportion of amictics and mictics are 43 and 57%, although the highest mictics in nature would be around 50% (Snell, 1987). In fact, Gilbert and Schroder (2004) have found that *Brachionus calyciflorus* mictic females began to appear from the fifth generation and increased to 50% by the 12th generation and thereafter, leveled off up to 18th generation.

Chemical signals produced by rotifer themselves are known to trigger mixis in four species of *Brachionus*, two species of *Epiphanes* and *Rhinoglena frontalis* (Wallace and Snell, 2010). Not surprisingly, *Brachionus* females newly hatched from resting eggs show extremely low propensity for mixis induction by the mixis-inducing factor. After a few parthenogenic generations, the different clones reach their maximum propensity for successful mixis induction (Gilbert, 2002). For induction of mixis, population threshold size ranges from 30 female/l to 250 female/l (Gilbert and Diegues, 2010). However, the mixis is induced even after one generation at the lowest density in *Hexarthra* sp inhabiting extremely unpredictable rock pool habitat (Schroder et al., 2007). In the monogonont rotifers, the chemical signal that induces mixis, appears and reaches the threshold level at high population density. Using appropriate techniques, Snell et al. (2006) discovered the chemical nature of the Mixis Inducing Protein (MIP) of ~ 39 KDa. The MIP is degraded by proteases but protected from degradation by natural protease inhibitors. On mixing proteinase K to water conditioned by one *B. plicatilis*/ml, the ability of mixis is reduced to 1%, where 17% neonates become mictic. Mixis is also triggered in *B. calyciflorus* by its own QS protein (see Wallace and Snell, 2010).

Another chemical that induces mixis is α tocopherol, the vitamin E. Carnivorous rotifers acquire tocopherol from preys like rotifers and others. Tocopherol is readily assimilated and > 60% of labeled tocopherol is transmitted from mothers to embryos. Incidentally, vitamin E is essential for fertility of male but not for female. The non-feeding male has to acquire it only from his mother. Being a direct inducer, it causes cytoplasmic enlargement during prenatal and postnatal development (Gilbert et al., 1979), increases the number of divisions in a variety of tissues (Wurdak and Gilbert, 1976), and nuclear size and level of DNA synthesis by endoduplication in some organs (Jones and Gilbert, 1977). However, no field level demonstration has been made to show that tocopherol induces mixis.

From data assembled for fecundity of amictic and mictic females in Table 9.3, the following may be inferred: (1) Between 16°C and 27°C, fecundity

TABLE 9.3

Effect of temperature on proportion of amictic, unfertilized and fertilized offspring in some rotifers

Species	Temperature (°C)	Amictic	Mictic		Reference
			Unfertilized	Fertilized	
Epiphanes senta	16–17	45.4	42.5	10.0	Ferris (1932)
Euchalanis dilatata	19	3.3	8.5	2.3	King (1970)
Ep. brachionus	20	31.9	28.3	7.4	Pourriot and Rieunier (1973)
Brachionus calycifilorus	20	29.8	24.8	3.3	Pourriot and Rieunier (1973)
Asplanchna brightwelli	20	10.0	14.8	4.1	Pourriot and Rieunier (1973)
B. urceslaris	20	20.0	25.0	4.0	Ruttner-Kolisko (1969)
A. sieboldi	22	-	14.6	6.1	Buchner et al. (1967)
Lecane inermis	19.24	20.7	14.2	-	Miller (1931)
Notommata copeus	23	34.5	31.9	-	Pourriot and Rieunier (1973)
Eu. dilatata	27	-	19.3	4.7	King (1970)
Total		195.6	223.9	41.9	
Average		24.45	22.4	5.24	
Amictic/mictic (%)		47	53		
Ratio between unfertilized and fertilized mictic eggs			4 : 1		

of mictics, unfertilized and fertilized mictics decreases with increasing temperature (Fig. 9.8A). This is unexpected, as larger eggs are produced at lower temperatures (Stelzer, 2002) and with larger egg size, the number of eggs is expected to decrease. It is not clear whether low temperatures facilitate increase in realized fecundity. More research is required before a generalization can be made. (2) Fecundity averages to 24.5, 22.4 and 5.2 eggs in amictics, unfertilized and fertilized mictics, respectively, i.e. the proportion of eggs produced by a female is 47% for the amictics and 53% for the mictics. (3) Of 27.6 mictic eggs, 22.4 and 5.2 eggs are unfertilized and fertilized, respectively, i.e. for every fertilized resting eggs, 4 unfertilized eggs are produced. Rearing *Brachionus calyciflorus* at four different algal (*Cryptomonas erosa*) densities, Gilbert (2010) reported that the lifetime fecundity of fertilized mictic females is increased from 0.9 resting eggs at 1.25×10^3 C. *erosa* cell/ml to 3.4 resting eggs at 2.5×10^4 cell/ml. Production of fertilized mictic eggs demands a larger input of reserves in the form of glycogen, lipid and cell secreting granules (Gilbert, 1993). Some 40–60% of the resting eggs are abnormal, especially in those that are produced at the age of 6 days and after in rotifers receiving higher or low doses of algal food (Fig. 9.8B, C). As hatching remains at around 40–60%, it is likely that all the normal resting eggs are hatched successfully. There are several types of abnormal eggs. The most common type is the egg shell filled with granular embryo mass leaving no extra-embryonic space. In others, the heterogenous embryo mass is ruptured. Hence, it is not clear whether the abnormality in eggs is due to reduced input of the reserves and/or cytological incompatibility.

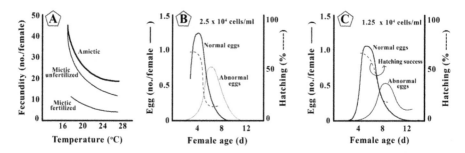

FIGURE 9.8

A. Fecundity of amictic, and unfertilized and fertilized mictic monogonont females (trends are drawn from data reported in Table 9.3). Normal and abnormal egg production and hatching success *Brachionus calyciflorus* fed on *Cryptomonas erosa* at B. high and C. low algal densities (compiled and redrawn from Gilbert, 2010).

Starvation and Rejuvenization

Rotifers often encounter limitations of food availability. To study the life history adaptation to starvation, many authors have deprived the rotifers from providing green alga at different ages. In fact, it is extremely difficult to achieve complete starvation in any aquatic animal. The alga-deprived rotifer species may still acquire bacteria and other microbes. They may also obtain dissolved organic substances like sugars and amino acids from surrounding waters (see Pandian, 2016). Regarding the so called 'starvation', the response of mixis pre-determined *Brachionus* species may differ from *Asplanchna* species, in which mixis determination remains labile for a relatively longer embryonic duration. Following starvation, *B. caudatus*, for example, ceased egg production without affecting survival, but *A. silvestrii* continued egg production at the cost of reduced survival (Fig. 9.9A, B).

In rotifers, egg production is an age-dependent activity (see Figs. 9.7B, C, 9.8B, C). In fact, many rotifers undergo aging and senescence. Neonates arising from old mothers have a shorter life span; their fecundity shows wider variability. Through a series of publications, Lansing showed that in *Philodina citrine*, maternal age influenced longevity and fecundity of F_1 neonates. Analyzing his publications, King (1983) reported that the short-lived lines appearing from old mothers reproduced earlier at higher rates than those arising from young mothers (see Wallace and Snell, 2010). The age and duration of starvation have profound effects on fertility and fecundity. In *B. plicatilis*, Yoshinaga et al. (2003) found that the initiation age and duration of starvation reduced (i) the proportion of reproductive females, (ii) reproductive period and (iii) thereby fecundity (Table 9.4). Surprisingly, one day starvation at the initial age of 2 days extended survival from

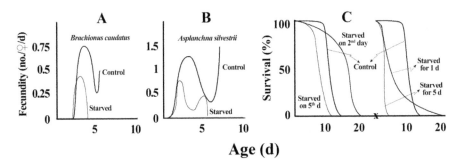

FIGURE 9.9

Age-specific fecundity of A. *Brachionus caudatus* and B. *Asplanchna silvestrii* starved for one day at the age of 1st day (compiled from Kirk, 1997). Survival of *Brachionus plicatilis* starved (C-left panel) for a day at the age of 2nd d or 5th d and (C-right panel) starved for 1 day or 5 days from birth and fed subsequently (compiled from Yoshinaga et al., 2003).

TABLE 9.4

Life history parameters of *Brachionus plicatilis* starved till different ages after birth (modified from Yoshinaga et al., 2003)

Age at First Feeding (d)	Reproductive Rotifer (%)	Lifetime Fecundity (no./l)	Reproductive Period (d)
0	100	25.7	9.42
1	75	16.4	9.25
2	50	10.1	8.00
3	25	4.8	4.25
4	0	0	0

~ 13 days to 20 days but starvation on the 5th day reduced it to 12 days, in comparison to 15 days in the control (Fig. 9.9C). From another experiment, they reported that during the post-starvation period, fecundity progressively decreased from 4.7 eggs, when starved at the age of 1 day to 2.2 eggs, when starved after the age of 4 days. Contrastingly, prior to starvation, it rapidly increased from 3.9 eggs in rotifers starved at the age of 2 days to 12.3 eggs in those starved at the age of 4 days. The starvation extended longevity from 4 days to 20 days. In *A. brightwelli*, reduced food intake extended life span by 14% and intermittent feeding extended it by 12% (see Wallace and Snell, 2010). It appears that the short period of starvation rejuvenates and extends the life span of rotifers recalling those reported for turbellarians (Pandian 2020).

Rotiferan Pseudocolonies

A vast majority of rotifers are solitary but ~ 25 species in eight genera and two monogonont families (Conochilidae, Flosculariidae) form what may be called pseudocolony. According to Blackstone and Jasker (2003), the colony is an association, in which the constituent members are intimately and organically connected together. The members in a rotifer 'colony' are not intimately connected, as are in bryozoan colonies (Wallace and Snell, 2010). In rotifers, the association may be named pseudocolony. Most (70%, 18 species) of these pseudocolonies are sessile. Some of them produce tubes from hardened secretions (*Limneas*) or gelatinous secretions (*Conochilus*) or pseudofecal pellets (*Floscularis conifera*). These secretions provide the structure and a substratum or matrix for the addition of new members to the pseudocolony. The number of members ranges from five in *F. ringens* to > 1,000 in *Lacinularia*.

On the basis of genetic relatedness, pseudocolonies can be formed by one of the following three methods: (i) In *Conochilus unicornis* (Fig. 9.10A), the

A **B** **C**

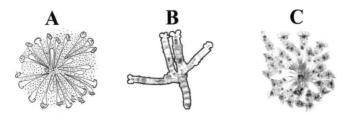

FIGURE 9.10

Three types of pseudocolony formation in rotifers. A. Autorecruitive *Conochilus unicornis* (from Hyman, 1951b), B. Allorecruitive *Floscularia conifera* (free hand drawing from Wallace and Snell, 2010) and C. Germinative *Sinantherina socialis*, dark spots indicate the brooded resting eggs (from Cladocera.de, all are free hand drawings).

recruitment can be auto-recruitive, as members are drawn from juveniles appearing from the same colony, (ii) it can be allo-recruitive, when a new member from an another pseudocolony is recruited, as in *Floscularia conifera* (Fig. 9.10B) and (iii) planktonic cohort of juveniles can form a new colony, as in *Sinantherina socialis* (Fig. 9.10C). Genetic relatedness decreases from auto-recruitive to other modes of recruitments (Wallace and Snell, 2010). In pseudocolonies of *S. socialis* with 11 and 32 members, 33 and 27% of females were carrying resting eggs and each female carried up to 5 resting eggs of different ages (Fig. 9.10C). Describing the mating behavior of pseudocolonial amphoteric sessile rotifer *S. socialis*, Rico-Martinez and Walsh (2013) indicated that the small male can mate with females of a size ranging from its own size to 8 times larger than its size.

Dormancy and Resting Eggs

To tide over unfavorable conditions, rotifers have devised dormancy in bdelloids and resting eggs in monogononts. In the wholly parthenogenic bdelloids, the distribution is limited to more permanent aquatic habitats, moist soil and mosses. In them, dormancy refers to the ability of adults to withstand and survive in a frozen or desiccated state. Some amazing examples for the ability of dormant adult bdelloids to tolerate low temperature and freezing or high temperature and desiccation are listed below: (i) *Philodina gregaria* resist repeated freezing and thawing. (ii) Soil inhabiting species like *Eucentrum mustela* cannot tolerate –8°C (see Gilbert, 1993) but a number of moss-dwelling bdelloids, in their frozen state, can withstand –4°C for a week and can endure supercooling down to –253°C. (iii) Whereas active *Philodina roseola* inhabits hot springs at 46°C (Hyman, 1951b), some calidinids survive on dry mosses exposed to 110–151°C for 35 minutes. Rotifers enter anhydrobiosis in stages. They contract first into a compact shape known as tun; subsequently, lose water and reduce their volume to 60% or even less. After drying in air for 2.5 years, *Macrotrachela quadricornifera* can be revived, when wetted (see Birky and Gilbert, 1971). After remaining dry and dormant in a dry herbarium for 59 years, *Callidina quadricornia* was revived after wetting (Rahm, 1923).

In contrast, monogononts, distributed from ephemeral rock pools holding water for < 7 days from the Chihuahuan desert (Schroder et al., 2007) to permanent lakes, have devised resting eggs. In the resting eggs, diapause is induced between the first few cleavages. Perhaps, the most interesting variability in resting eggs is that of *Brachionus calyciflorus*, in which female lays large (150 µm) orange and smaller (140 µm) gray colored eggs. *B. diversicornis* also produce dark brown and light colored eggs (see Gilbert, 1971).

In some rotifers, the resting eggs have spines, hairy filaments or pin-like extensions (*Ploesoma, Synchaeta*) or are filled with oil droplets (*Asplanchna*) or air bubbles (*Filinia*), which may help them float and facilitate their spatial distribution. The outer surface of the sediment settled resting eggs are variedly decorated with projections, furrows and folds (Fig. 9.11). In crustaceans, fertilized resting eggs from successive episodes of sexual reproduction accumulate in sediments and can remain dormant for years. The accumulation of resting eggs results in the formation of an 'egg bank', and their sporadic hatchings after different durations facilitate temporal distribution and sustenance of the population. In seven rotifer species, the accumulated resting eggs density ranges from 100,000/m² to 4,000,000/m². These values may be compared with 350 to 700,000 cyst/m² for the entomostrucan crustaceans and 20,000 to 28,000 resting egg/m² for the daphnids (Pandian, 2016). From sediments, resting eggs remain hatchable up to 22 years in *Polyarthra dolichoptera* and 40 years in others. Resting eggs of *Daphnia pulicaria* and cysts of the copepod *Diaptomus sanguineous* can be hatched after 125 years and 322 years, respectively (Pandian, 2016). Hence, the capacity of rotifers to accumulate resting eggs in sediments and to hatch them subsequently is less than that of crustaceans.

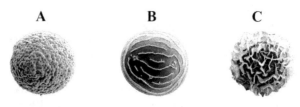

A   B   C

FIGURE 9.11

Resting eggs of A. Asplanchna, B. Brachionus and C. Kellicottia (permission by Dr. H. Segers, Royal Belgian Institute of Natural Museum).

There are succinct differences between 'dry' crustacean cysts and 'hydrated' resting eggs of daphnids and rotifers. For example, the *Artemia* cyst contains very little water but the resting eggs of rotifers hold ~ 70% water (Hagiwara, 1996). As in cysts, the resting eggs of *Brachionus plicatilis* also produce and store abundant amounts of chaperons, antioxidants and Late Embryogenesis Abundant (LEA) proteins. However, it is not yet known whether it also suppresses metabolism, as in *Artemia* cyst (Clark et al., 2012). Investigations into the hydrated embryos of rotifers and cladocerans may lead to developing methods for long term preservation of human cells and human itself.

Sex in Bdelloids?

In bdelloid rotifers, the presence of male or male sex organ(s), however, have never been observed (Birky, 2010). Their oocytes are formed through mitotic division with no reduction of the chromosome number and no indication of chromosome pairing (Flot et al., 2013). With inbreeding depression and accumulation of deleterious genes, strict clonal reproduction must lead to an evolutionary dead end (Debortoli et al., 2016). Nevertheless, the bdelloids have persisted and diversified into 461 species during the last 60 million years. This has posed a challenge but attempts have been made in the last few years to know whether they have alternative means of gaining genetic diversity.

Bdelloids are known for their remarkable ability to survive desiccation. In them, the heat shock-gene *hsp*82 has been characterized. In *Philodina roseola*, there are four copies of *hsp*82 but each is located on a separate chromosome. Welch et al. (2008) sequenced 45–70 kb regions surrounding the four *hsp*82 copies. Their arrangement suggests that the genome is a degenerate tetraploid. Confirming it in *P. roseola* and *Adineta vaga* belonging to two different bdelloid families, Hur et al. (2009) added that the degenerate tetraploidy was established prior to the two families being diverged and therefore, probably before bdelloid radiation. It has also been shown that the highly diverse gene copies in the degenerated tetraploids are homeologs rather than homologs and their genes evidence for a sort of 'sexuality' (Welch et al., 2008, Hur et al., 2009). Incidentally, unlike other clades like tardigrades, bdelloids tolerate desiccation by not producing trehalase, a non-reducing sugar that is considered to contribute to the production of cellular constituents against desiccation damage (Gladyshev and Meselson, 2008). Further, the draft sequence of genome of *A. vaga* revealed a peculiar genomic structure incompatible with meiosis, indicating that DNA damages, in the form of Double-Strand Breaks (DSBs), are gradually repaired upon rehydration. In fact, Hespeels et al. (2014) found that *A. vaga* individuals were able to efficiently recover from desiccation and repair a large amount of DSBs.

Flot et al. (2013) noted that at the gene scale, the *A. vaga* genome is tetraploid and is composed of both anciently duplicated segments and less divergent allelic regions. They also provided ample evidence for gene conversion, which may limit the accumulation of deleterious mutations in the absence of meiosis. Over 8% of their genes are likely to be of non-metazoan origin and are acquired from conversion of genes during DNA repair following the damages due to desiccation (Hespeels et al., 2014). The haplowebs and GMYC analyses of 576 individuals of *A. vaga* by Debortoli et al. (2016) revealed the existence of six species among the samples with strong evidence of both intra- and inter-specific recombinations. Eyres et al. (2015) demonstrated that the level of horizontal gene transfer is higher in the bdelloid *Rotatoria* species that experience regular desiccation events in their semi-terrestrial habitats. Adopting population genetics, which provide definitive means to detect infrequent or atypical sex, Signorovitch et al. (2015) found that segregation and allelic sharing occur without requiring homologous chromosome pairs in the mitochondrial clade of the bdelloid *Macrotrachela quadricornifera*. The foreign genes are accumulated in telomeric regions of the chromosome. Briefly, it appears that the bdelloids have attempted to have 'sexuality' by means other than the conventional meiosis and recombination.

10

Gastrotricha

Introduction

The gastrotrichs are free-living, elongate, ventrally flattened, microscopic (0.1 to 1.5 mm in length, Hyman, 1951b), dorsally covered with cuticular scales, spines and/or bristles, pseudocoelomate eutelic (Hummon, 1984) hairy bellied worms. Their head lobe and ventral surface are ciliated. In them, the sub-terminal mouth is followed by muscular suctorial pharynx with a Y-shaped lumen, and a simple gut terminating in an anus. The muscular system consists of muscle bands (but long layers) in four orientations: circular, longitudinal, helicoidal and dorsoentral. There is a pair of protonephridia but no circulatory or respiratory system. The brain is bilobed and is followed by a pair of lateral nerve cords.

Taxonomy and Distribution

The gastrotrichs comprise 813 species in 17 families and 2 orders: Macroasyida and Chaetonotida (Balsamo et al., 2015, Animal Diversity Web). The Macroasyida are distinguished from the Chaetonotida by the presence of pharyngeal pores and two pairs of adhesive tubes (Fig. 10.1). The former comprise 300 species in ~ 25 genera and 6 families. The latter have > 400 species in 22 genera and 7 families (Hummon, 1974, Animal Diversity Web). All macrodasyids and members of two chaetonotid families Neodasyidae and Xenotrichulidae are marine gastrotrichs. All the remaining chaetonotids are freshwater and limnoterrestrial species. Approximately, 53 and 47% of gastrotrichs are marine and freshwater habitants, respectively (Animal Diversity Web). They are abundant and occur in densities of million individual/m². They may constitute 1–8% of benthic meiofaunal organisms in marine waters. In freshwaters, they are benthic or deriphytic.

As mentioned earlier, gastrotrichs are one among the few animals commonly found in anaerobic sediments. It is likely that they possess a

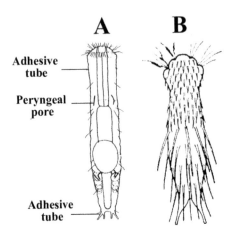

FIGURE 10.1

A. Macroasyidan, *Redudasys formerise* and B. chatonoitid *Chaetonota* (free hand drawing from Strayer et al., 2010).

sulfate detoxification mechanism. However, their density, for example, decreases from ~ 1.8 million/m² in the littoral area of Lake Brzeiziczno, Poland to almost zero at 16–17 m depth (Nesteruk, 1991). In the lakes of Poland and Italy, their biomass is in the range of 23–30 mg dry weight/m² (Strayer et al., 2010).

Reproductive Biology

In gastrotrichs, reproduction commences with parthenogenic (2-6 ovipositions, e.g. *Aspidophorus polystictos*, Table 10.1) four ovipositions (Hummon, 1986). Subsequently, it is switched to hermaphroditic reproduction. Some gastrotrichs like *Dactylopodola* are protandrics and may have males or females alone. The female reproductive system includes single (in Lepidodasyidae, Thaumastodermatidae) or paired ovaries with clusters of distinct oogonia, oviduct, seminal receptacle, copulatory bursa, female antrum, paired glands and gonopore. Ovaries are located anterior to the oviduct in most gastrotrichs (e.g. *Macrodasys*, Fig. 10.2B, D) but posterior to it in the macrodasyid families Leptodasyidae and Planodasyidae and chaetonotid family Neodasyidae (e.g. *Cephalodasys*, Fig. 10.2A, C). When all the component organs are present, they lay in linear series with the sperm penetrating from the gonopore through the bursa and receptacle to the oviduct; the eggs are spawned in reverse order from the oviduct to the gonopore or simply released by rupture

TABLE 10.1

Effect of temperature and salinity on life history traits and reproduction in chaetonotid *Aspidophorus polystictos* (compiled and modified from Balsamo and Todaro, 1988)

Traits	Temperature (°C)		Salinity (‰)		
	20	25	15	35	55
Life span (d)	22.6	23.2	24.9	23.2	11.0
Embryonic (d)	2.0	1.7	1.4	1.7	1.8
Pre-reproductive (d) as % of life span (%)	4.2 18.5	2.7 11.6	2.1 8.4	2.7 11.6	4.5 40.9
Parthenogenic (d) as % of life span	8.7 38.5	7.1 30.7	4.8 19.3	7.1 30.6	3.0 27.3
Hermaphroditic (d) as % of life span	9.2 40.7	13.4 58.0	18.0 72.3	13.4 57.8	3.7 38.8
Oviposition (no.)	6.0	4.0	4.0	5.0	5.0
Parthenogenic fecundity (no./♀)	4.6	5.0	1.0	4.0	4.0
Hermaphroditic fecundity (no./♀)	2.0	-	-	-	-
Doubling time (d)	3.4	2.2	1.2	2.2	5.8
Reproductive effort	0.5	0.8	1.3	0.8	0.7

of the body, as in most cases. The male reproductive system consists of an elongate pair of testes, vas deferens and male gonopore (Hummon, 1974).

In the majority of gastrotrichs, parthenogenesis is the only mode of reproduction, especially in members of four families, in which either the testes remain rudimentary (e.g. Chaetonotidae, Fig. 10.2G) or completely absent in members of Neogoseidae, Dichaeturidae, Proichthydidae (Hummon, 1974) (Fig. 10.3 left cycle alone). In *Lepidodermella squamata*, successive parthenogenic eggs originate from alternate ovaries (Hummon, 1984). The first three eggs are the tachyblastic subitaneous eggs that develop and hatch quickly within a day after laying at 20°C. The fourth and last egg is usually but not always the opsiblastic resting egg and is produced sporadically. These resting eggs are resistant to freezing and drying (Strayer et al., 2010). For example, of 465 animals reared, only 125 or 27% produced the resting eggs, which did not hatch until the seventh month of observation. The changes in the temperatures and food quality did not alter the parthenogenic fecundity of four eggs, albeit inter-oviposition duration was extended at 20°C (Fig. 10.4). However, these changes extended survival from 23 days at 27°C to 37–48 days at 20°C (Hummon, 1986).

However, in many gastrotrichs, the Life Span (LS) includes a fairly long post-parthenogenic hermaphroditic phase from 39 to 72% of LS (Table 10.1). In the marine gastrotrich *Aspidophorus polystictos*, the

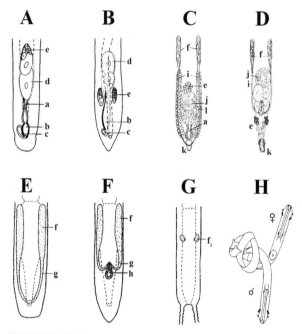

FIGURE 10.2

Reproductive system of females: A. *Cephalodasys*, B. *Macrodasys*, hermaphrodites C. *Cephalodasys*, D. *Macrodasys* and males E. *Cephalodasys*, F. *Macrodasys*, G. *Polymerurus*. H. Copulation in *Turbanella cornuta*. a = seminal receptacle, b = copulatory bursa, c = feminine antrum, d = large oocyte in oviduct, e = oocyte proliferation site of the ovary, f = testis, f1 = degenerated testis, g = vas deferens, h = penis, i = sperm ducts, j = uterus containing ripe eggs, k = female gonopore, l = yolk gland (redrawn after Remane, 1936).

parthenogenic phase lasts for 19% (or 4.8 days) and 31% (or 7.1 days) of LS at 15 and 35‰, respectively; it lasts for 31 (or 7.1 d) and 39% (or 8.7 d) at 25°C and 20°C. These are 4 days for parthenogenic phase and 6 days for post-parthenogenic phase at 20°C in the freshwater gastrotrich *Chaetonotus maximus*, the life span lasts for 4.4 days at 20°C and 25°C (Balsamo and Todaro, 1988). During the post-parthenogenic phase, these gastrotrichs switch to hermaphroditism. *Lepidodermella squamata* sperms were identified (Weiss and Levy, 1979) and were found grouped in a median single or bilateral pair of testicular sacs within the testes. Each sac contains 16 sperms only, i.e. the number of spermatocytes is limited to four only. The sperm is fairly long (~ 150 μm in length in animals with body length of 240 μm, Hummon, 1974) and non-motile. The sperms are encomposed in a spermatophore, which is attached to the mating partner. In some of them, mating and copulation are described by Teuchart (1968). However, the hermaphrodites bear flagellated

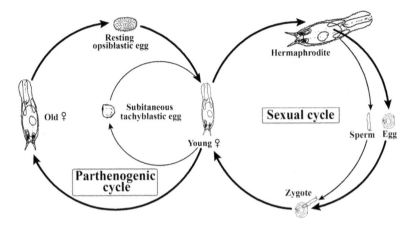

FIGURE 10.3

Life cycle in gastrotrichs. In some life cycle, reproduction is limited to parthenogenic cycle alone. In others like *Lepidodermella squammata* (based on Hummon, 1986, Strayer et al., 2010), the heterogonic cycle involves initially by parthenogenic and subsequently by sexual reproduction.

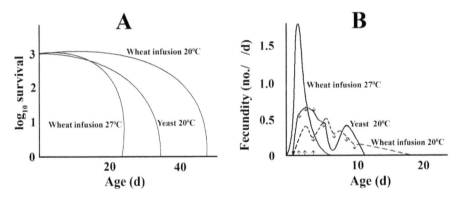

FIGURE 10.4

Effect of temperature and food quality on A. survival and B. parthenogenic fecundity in *Lepidodermella squammata*. Irrespective of changes in temperature and food quality, the gastrotrich lays only 4 parthenogenic eggs and are indicated by arrow heads (redrawn from Hummon, 1986).

sperm in Macrodasyidea (Hummon, 1986). The embryonic development is indicative of a modified radial cleavage. For more details, Hummon (1974) and Hejnol (2015c) may be consulted.

Balsamo and Todaro (1988) is perhaps the only publication, which provides information on the effect of temperature and salinity on the life history traits, fecundity and reproductive effort. They have figuratively reported the number of parthenogenics and sexual fecundity. Only at temperature 20°C, hermaphrodites produced 2 egg/♀/d (Table 10.1); at all other temperatures and salinities, hermaphrodites failed to produce eggs. The other reported values are recalculated and processed to draw some inferences: (1) At 20°C and 25°C as well as 15‰ and 35‰ salinities, the life span is ~ 24 days; only at super-salinity of 55‰, it is reduced to 11 days. (2) At 15 and 35‰ salinity and at 25°C, the animal grew faster, as the doubling time is ~ 2 days. (3) Whereas parthenogenic fecundity is in the range of 4–5 egg/♀, irrespective of changes in survival, food quality and temperature (Fig. 10.4), the hermaphroditic fecundity is around 0.2 egg/♀ at 25°C. (4) Reproductive effort is < 1 at these salinities and temperatures, as *A. polystictos* reproduce predominantly by parthenogenesis.

11

Kinorhyncha

Introduction

The 200 speciose (Hejnol, 2015a) Kinorhyncha are free-living microscopic, elongate (1 mm in length) bristly or spiny pseudocoelomate worm devoid of cilia. Their body is covered with a thick cuticle and is superficially segmented by 13 joints called the zonites. Their spherical head constitutes the first zonite and bears a terminal mouth surrounded by five to seven circlets of posteriorly directed spines, the scalids. The head is withdrawable up to the second or third zonite. Their other systems resemble those of rotifers and gastrotrichs (Hyman, 1951b). The kinorhynchids are divided into two suborders: Cyclorhaghae and Homalorhagae (Neuhaus and Higgins, 2002). They are exclusively marine inhabitants, occur below oxygenated layers and are recorded from 5,300 m depth in the Pacific (Meadow et al., 1994). Rao and Satapathy (1996) found that *Echinoderes* and *Pycnophye* survive salinities of 5.7–20.3‰, pH 7.64–9.50 and temperatures up to 33.5°C in an Indian lagoon.

Reproductive Biology

The kinorhynchs are gonochores but sexes are usually not distinguishable externally. A pair of saccate gonads opens separately on the 13th zonite. The gonad consists of an anterior apical cell that gives rise to all other cells of it. Initially, the ovary is syncytial and consists of germ cells, nutritive cells (in females only) and a surrounding epithelium. Subsequently, an ovum matures drawing nutrients from the nutritive syncytium (Hyman, 1951b). In the center of the ovary, a single oocyte is developed containing numerous vesicles, enormously large nucleus and nucleolus (Fig. 11.1A). In *Echinoderes dujardinii*, the egg measures ~ 75 µm in size (Kozloff, 1972). The seminal receptacle is located at the posterior end.

From the testes, short gonoduct leads to a gonopore, which is armed with penial spicule (Hyman, 1951b). At the anterior end of the testis,

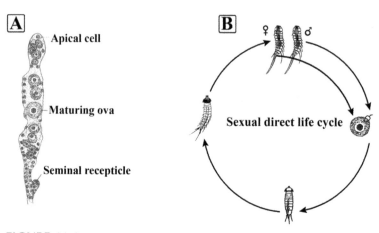

FIGURE 11.1

A. Echinoderid ovary (free hand drawing from Hyman, 1951b) and B. life cycle of a kinorhynch.

spermatogonial differentiation is commenced and the mature spermatocytes and spermatids are seen at the caudal gonadal part (Adrianov and Malakhov, 1999). The cigar-shaped mature sperm measures up to 25% of the worm body length. The large sperm consist of a broad head harboring mitochondria and rod-like nucleus and a short tail. As the vesicle of young female contains sperm, fertilization must be internal.

In *E. kozloffi*, the first few cleavages are equal and may be spiral. The juvenile hatches after 9 days at 9°C (Kozloff, 2007). The development is direct but involves five to six molts (Sorensen et al., 2010). At hatching, *E. brookhouti* is 11-segmented, and the 12th and 13th segments are added during the first and second molts, respectively (Fig. 11.1B, see also Higgins, 1974). The available publications do not indicate eutelism in kinorhynchs. It is also not clear whether growth achieved by juveniles during the molt represents a simple expansion of cytoplasm of the existing cells at hatching or cells are added during each molt. The less speciose kinorhynchs with fewer eggs indicates that they may be eutelic, as in gastrotrichs. Incidentally, the nematode hatchlings undergo four molts and six stages during development but they are also eutelic (Poinar, 2010).

12

Nematoda

Introduction

The nematodes are fusiform (e.g. free-living *Rhabditis elegans*, Fig. 12.1A, parasitic *Trichinella spiralis*, Fig. 12.1B), or filiform (e.g. *Nematodirus*, Fig. 12.1C) or partially fusiform and partially filiform (e.g. *Trichuris ovis*, Fig. 12.1D) or saccate (e.g. *Heterodera schachtii*, Fig. 12.1G) cylindrical worms with no cilia or flame bulbs. They have four main longitudinal epidermal chords, a three-angled pharynx, circumventric nerve ring and one or two tubular gonads opening separately in the female but into the rectum in males. The pharynx plays an important role as a pump by forcing food into the gut (Schiemer, 1987). Their body wall consists of an outer cuticle, a middle cellular or syncytial sub-cuticular epidermis and inner muscular layers of exclusively longitudinal fibers. They may be transparent, whitish or yellowish. Their size ranges from up to 50 mm in free-living marine worms to > 1 m in parasitic worms like the kidney worm *Dioctophyme renale* and the guinea worm *Dracunculus medinensis* (Hyman, 1951b) and a life span from 7 days in *Acroblepoides* sp to ~ 7 years (y) in *Necator americanus* (see Morand,

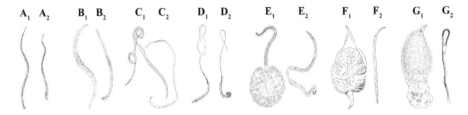

FIGURE 12.1

A_1, A_2. Fusiform female and male of free-living *Rhabditis elegans* (after Maupas, 1900), B_1, B_2. female and male of *Trichinella spiralis* (after Leuckart, 1876), C_1, C_2. filiform female and male of *Nematodirus* (after Ranson, 1911), D_1, D_2. female and male of *Trichuris ovis* (after Ranson, 1911), E_1, E_2. female and male of *Simondsia* (after Cobbold, 1883), F_1, F_2. female and male of *Tetrameres* (after Cram, 1931) and G_1, G_2. female and male of *Heterodera schachtii* (after Strubbel, 1888).

1996). In *Metastrongylus elongatus*, the dauer/dormant juvenile stage lasts for 4 years in earthworms.

Taxonomy and Distribution

The number of described nematode species ranges from 27,000 (Abebe et al., 2008) to 27,333 (see Bernard et al., 2017), i.e. the number increased at the rate of 37 species/y during the period from 2008 to 2017. Morand et al. (2005) reported that the number increased at the rate of 364 species/year during the period from 1979 to 1988 (Hammond, 1992). However, the estimated number ranges from one to ten million; at the present rate of identification and description, hundreds of years may be required to complete taxonomy of nematodes. A striking feature of nematodes is their species richness, numerical abundance and ecological omnipresence. For example, the soil nematodes occur in > 1 million worm/m^2 (Sommer, 2015). Their species richness is greatest between 30° and 40° latitudes (94 species/sample) but is decreased near the equator (0° to 10°, 81 species/sample, Boag and Yeates, 1998). Among free-living nematodes, 50, 33 and 17% of them are herbivores, bacterivores and fungivores, respectively (McSorby, website).

The phylum Nematoda was divided into two classes: Adeonophorea and Secernentea. The former was further divided into 7 orders and latter into 9 orders (Hyman, 1951b). It was later divided into two classes Enoplea with three subclasses and eight orders, and Dorylaimea with three subclasses and nineteen orders (Zhang, 2011). Based on molecular phylogeny, De Lay and Blaxter (2002) divided it into five major clades namely Dorylaimia (order: Trichurida), Enoplida, Spirurina (Orders; Ascaridida, Oxyurida, Spirurida), Tylenchina (order: Rhabdiasoidea) and Rhabditina (orders: Rhabditida, Strongylida). As more relevant information on reproduction and development is provided, Hyman's classification is presented for an easier understanding (Table 12.1).

Poinar (2010) assembled interesting information on nematode distribution, abundance and production in freshwaters. Accordingly, the nematodes constitute 60% of all benthic metazoans in Lake Mirror, USA. While only 18% of them penetrate to a depth deeper than 4 cm, 60% of them are restricted to 2 cm depth. However, *Eudorylaimus andrassyi* is reported from 43 m depth in Lake Tibrias, Isreal. In Neusiedlersee, Austria, *Tobrilus gracilis*, *Ethmolaimus* and *Monhystera* occur and thrive in muddy anoxic zones. Due to differences in size, the limnic nematodes constitute 1–15% or 40–80% of zoobenthic biomass. In New Zealand alpine streams, nematodes constitute 40.6% of the moss-dwelling invertebrates. In other lentic waters, they make up 6.4 to 14.6%, i.e. 38, 350 worm/l (Glime, 2017c). Known for their abundance, they undergo remarkable seasonal change. In an estuary, England, the density

TABLE 12.1

Classification of nematodes (Hyman, 1951b). Names of orders are shown in bold letters and some representative families in normal letters. M = marine, FW = freshwater, T = terrestrial habitants, P = parasite

Enoploidea M + FW sediment Enoploidae	**Dorylaimidea** FW + T sediments Dorylaimidae	**Mermithoidea**, ℗ Tetradonematidae Mermithidae	**Chromadoroidea**, M Cyatholaimidae Choanolaimidae
Araeolaimoidea M + FW sediment Axonolainidae Camacolainidae	**Monhysteroidea** M + FW sediments Limhomoeidae	**Rhabditoidea**, ℗ FW sediment Rhabditidae Cephalobidae Steinernematidae Tylenchidae Atlantonematidae	**Rhabdiasoidea**, ℗ Rhabdiasidae Strongyloididae
Oxyruoidea, ℗ Oxyruidae Atractidae Kathlaniidae Heterakidae Subuluridae	**Ascaroidea**, ℗ Ascaridae Anisakidae Goeziidae	**Strongyloidea**, ℗ Stronglylidae Syngamidae Ancylostomidae Heligmosomidae Trichostrongylidae Metastrongylidae	**Spiruroidea**, ℗ Spiruridae Thelaziidae Acuariidae Gnathostomidae Camallanidae Cucullanidae
Dracunculoidea, ℗ Dracunculidae	**Filarioidea**, ℗ Filariidae Setariidae Aproctidae	**Trichuroidea**, ℗ Trichuridae Trichosomoididae Trichinellidae	**Dioctophymoidea**, ℗ Dioctophymidae Soboliphymidae

of 40 nematode species changes from 23 × 10^6/m^2 during late spring to ~ 8.5 × 10^6/m^2 during winter but that of *T. grandipapellatus* from 2,35,000/m^2 in winter to 6,00,000/m^2 in summer. An estimate of the production indicates 6.6 g C/m^2/y. *Caenorhabditis briggsae* converts bacteria into nematode tissues at the efficiency of 13–20%; in other nematodes, the efficiency ranges from 5 to 25% (Schiemer, 1987).

Reproductive System

The nematodes are gonochores. Males can readily be distinguished externally from females by their smaller size, curvature of the posterior end and presence of bursae, genital papillae and other accessory copulatory structures (Fig. 12.1A–D). In *Simondsia*, *Tetrameres* and *Heterodera schachtii*, the level of dimorphism is so high that the females grow to attain a lemon shaped body, while their males retain the original shape (Fig. 12.1E, F, G). The nematode gonad is tubular in shape, may be straight (Fig. 12.2A), reflexed (Fig. 12.2B) and didelphic with anterior parallel ovaries (Fig. 12.2C, E) or monodelphic sinous ovary (Fig. 12.2D). In the telogonic gonads, germ

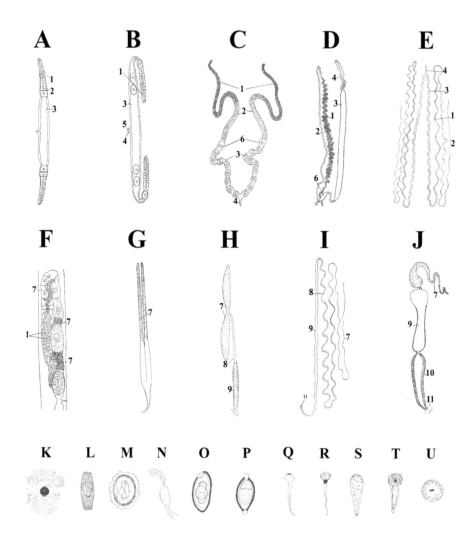

FIGURE 12.2

Schematic didelphic female reproductive system with (A) straight ovaries, (B) opposite flexed ovaries (free hand drawings from Hyman, 1951b), (C) anterior parallel ovaries in *Heterodera* (after Rauther, 1918), (D) monodelphic sinous ovary in *Trichuris*, (E) parallelly long didelphic system in *Ascaris* (free hand drawings from Hyman, 1951b). (F) Ovotestis showing alternating locations of testis and ovary in a hermaphrodite (from Potts, 1910). Diorchic male reproductive system in (G) *Heterodera marioni* (after Atkinson, 1889), (H) with two opposite testes. (I) Monorchic male reproductive system in *Ascaris* (free hand drawings from Hyman, 1951b), (J) with an anterior testis in *Camallanus* (after Tornquist, 1931). Examples are shown for eggs and sperm. Note that presence or absence of flagellum in some sperm. (K) Mermithid egg with complex filaments (after Christie, 1937), (L) spiral filaments enclosing *Pseudonymus* egg (after Gyory, 1856), (M) *Ascaris* egg, (N) *Tetrameres* egg with polar filaments, (O) oxyurid egg with a single lid (after Seurat, 1920) and (P) trichuroid egg with double lids (after Alicata, 1935). 1 = ovary, 2 = oviduct, 3 = uterus, 4 = vagina, 5 = vulva, 6 = seminal receptacle, 7 = testis, 8 = sperm duct, 9 = seminal vesicle, 10 = ejaculatory duct, 11 = intestine. (Q-U) Sperm types in nematodes.

cells are proliferated only at the blind proximal end, the germinal zone (e.g. *Ascaris suis*) or through the entire length of the gonad in hologonic gonads (e.g. *Heterodera*). In females, the system consists of a single (monodelphic, e.g. Enoploidea) or two (didelphic, e.g. Mermithoidea) tubular ovaries, each continuing with a separate oviduct to the exterior through a widened uterus, muscularized vagina and vulva (Fig. 12.2A–E). Fertilization occurs in the uterus, where the eggs and sperm are stored; hence, the uterus serves as seminal receptacle/spermatheca. In males, the system consists of one (monorchic, e.g. Secernentea) or two (diorchic, e.g. Enoploidea) tubular testes, which continue as sperm ducts to converge in a common vas deferens or ejaculatory duct (Fig. 12.2G–J). The duct opens into the rectum (Hope, 1974). In protandric hermaphrodites, testes and ovaries are arranged alternatively (Fig. 12.2F).

In the absence of a flagellum, the nematode sperms are non-motile (Fig. 12.2Q–U) and released into the spermathecal area of the uterus, where eggs are stored. The dwarf male of *Trichosomoides crassicauda* (1.5–3.4 mm in length) dwells in the vagina of females (10–13 mm) to ensure fertilization by non-motile sperms. Immediately following fertilization, a fertilization membrane is formed and the membrane is then thickened to form the shell around the egg. As the egg passes through the uterus, a protein membrane is also formed, which furnishes the surface sculpturing, filaments and so forth. A single or a bunch of filaments (e.g. *Tetrameres*, Fig. 12.2N) is formed from the protein membrane; in mermithids, it is elaborately branched with projections, the byssus (Fig. 12.2K). A spirally coiled filament encircles the egg and is released on contact with water in *Pseudonymus* (Fig. 12.2L). These filaments aid in entangling and attaching the eggs to aquatic plants (Hyman, 1951b).

Segregated from the somatic cell lineage early in development, the paired primordial and epithelial cells may develop into one or two gonads. In the didelphics, the germ cell nuclei are gradually separated by developing gonoducts and each gonad receives a single Primordial Germ Cell (PGC). However, the primordial epithelial cells, located to the posterior of the germ cell, do not facilitate the separation of germ cells in the monodelphics. As a consequence, a single gonad is developed with two PGCs (Hope, 1974). According to Sommer (2015), the delphic number and structure of nematode gonads vary greatly between the two sexes and across species with male gonads being one armed (monodelphic) and female/hermaphrodite gonads being two armed (didelphic). In the hermaphroditic *Caenorhabditis elegans*, the two gonadal arms develop almost symmetrically from two somatic precursor cells Z1 and Z4. Cell lineage studies on the monodelphic *Panagrellus redivivus* have revealed that the monodelphy results from the programmed cell death of the posteriorly located Z4. Within the family Rhabditidae, monodelphy has evolved repeatedly. However, it is not clear whether the monodelphy appeared by the location of the PGC or programmed death of Z4 during different times of its origin.

Incidences of anomalous intersexes, i.e. females with spicula or testis with no sperm and males with precursor vulva or ovary with no eggs are listed by many authors (e.g. Hope, 1974). However, the description for hermaphrodite gonad strucuture is limited to an earlier publication by Potts (1910). This description indicates the presence of an undivided, undelimited single chambered (see also Fig. 7.2E) ovotestis, in which the ovarian and testicular zones are alternatively located (Fig. 12.2F). As self-fertilization seems to be the rule, the gametes may be released simultaneously. More importantly, the hermaphroditic species produce a limited number of sperm to fertilize only a fraction of the eggs so that a period of fertility is followed by a longer period of laying unfertilized eggs. At 20°C, the duration of fertilized and unfertilized eggs laying in *Ceanorhabditis elegans* lasts from 2 to 6 d and 5 to 10 d, respectively (Klass, 1977). Potts (1910) followed the sex ratio of progenies produced from the 1st to 19th generation in the hermaphrodite *Diplogaster maupasi*. The ratio was 0.96 ♀ : 0.04 ♂ during the first five generations but from the 6th to 19th generation, only hermaphrodites were produced. Hence, the chances for cross fertilization are limited in hermaphrodites. The hermaphrodites can mate with males or can also self-fertilize, especially in animal parasitic species that are isolated. In families Rhabditidae and Diplogasteridae, androdioecious, i.e. mating between hermaphrodite and male (Pandian, 2016) mating system has evolved multiple numbers of times independently (Denver et al., 2011).

Unlike most animal taxa, reproduction in many nematodes is often limited by the availability of males and sperms rather than females and oocytes. For example, the potato rot nematode *Ditylenchus destructor* male can transfer only 6–20 sperm per copulation (Anderson and Darling, 1964). In the protandric hermaphrodite *Caenorhabiditis elegans*, ~ 280 sperms are produced during the male phase. However, 1,000 oocytes are generated during the subsequent female phase (Kimble and Ward, 1988). Hence, each of 3.6 eggs awaits fertilization by a single sperm. As a consequence, production of infertile eggs is a common feature of nematodes. For example, 140–230 fertilized eggs are laid for 3 days but infertile eggs for 6–8 days in *Diplogaster robustus* with a life span of 16 days. In long living *Cephalobius dubius*, the duration of ovipositing fertile eggs lasts for 2 months and infertile eggs for one month or longer (Hyman, 1951b). Hence, there is a need to know whether the spermatogonia remain eutelic, while the oogonia are exempted from it.

Parthenogenesis

In parthenogenesis, mating and genetic recombination do not occur. However, meiotic parthenogenic eggs (e.g. *Rhabditis monhystera*, Hyman, 1951b) require activation by the sperm. The majority of parthenogenic nematodes reproduce by obligatory mitotic parthenogenesis. The following are some interesting examples for adoption of different parthenogenic types: (1) *Mesorhabditis belari* reproduce mostly by gynogenesis/pseudogamy but rarely sexually. (2) One race of *Meloidogyne hapla* or one population of *Pratylenchus scribneri* reproduce by meiotic parthenogenesis, while the others by mitotic parthenogenesis; for example, *P. scribneri* reproduce by meiotic in Florida and mitotic parthenogenesis in California, respectively (Cable, 1971). In freshwater and terrestrial nematodes, the preponderance of females has been reported. Of 45 species examined, 15 had no males. In 50% of freshwater and 75% of terrestrial species, sex ratio is skewed in favor of females (Hyman, 1951b). (3) In some of them, parthenogenesis was deduced from the laboratory and other observations. In the absence of males, *Heterotylenchus* reproduce parthenogenically but alternate with sexual reproduction (Cable, 1971). Reared in the laboratory, *Chromadorina epidemos* and *Viscosia macramphida* produce parthenogenic females alone, although they reproduce sexually in the fields (see Hope, 1974). The occurrence of parthenogenesis has been demonstrated in several terrestrial nematodes like *Mermis subnigrescens* and *Heterodera marioni*, albeit males are known for both species (Hyman, 1951b).

TABLE 12.2

Nematode species reported as parthenogenics

Obligatory Mitotic	Facultative Meiotic
Plant parasites: *Heterodera marioni* (Hyman, 1951b); *Melodoigyna africana, M. ardenensis, M. arenaria, M. enterolobii, M. ethiopica, M. hapla, M. hispanica, M. incognita, M. izalcoens, M. javanica, M. konaens, M. oryzae, M. paranaens, M. partityla, Pratylenchus brachyrus, P. neglectus, P. zeae* (Janssen et al., 2017); *Deladenus siricidicola* (Hajek and Eilenberg, 2018); **Animal parasite:** *Mermis subnigrenscens* (Hyman, 1951b) **Free living:** *Cephalobius dubius* (Welch and Wehrle, 1918)	**Plant parasites:** *Melodoigyna chitwoodi, M. exigua, M. fallax, M. floridensis, M. graminicola, M. graminis, M. hapla, M. minor, M. naasi* (Janssen et al., 2017); **Animal parasites:** *Strongyloides fuelleborni, S. papillosus, S. ransomi, S. ratti, S. simiae, S. stercoralis* (see Streit, 2008), *S. venezuelensis* (Hino et al., 2014); *Mesorhabditis belari, Pratylenchus scribneri, Heterotylenchus* (see Cable, 1971), *Chromadorina epidemos, Viscosia macramphida* (Hope, 1974); **Free living:** *Caenorhabditis elegans, Diploscapter lycostoma, D. pachys, Protorhabditis* (Schwarz, 2017)

Repeated computer search revealed the existence of ~ 45 parthenogenic nematode species (Table 12.2). Of them, ~ 20 and 25 are mitotic/apomictics and meiotic/automictic parthenogen, respectively. Most of them are plant (~ 28 species) or animal (~ 12 species) parasites. The latter belong to either Strongyloidae or Rhabditidae. The life history of some of these parthenogens is described in Fig. 12.12. In these nematodes, cytological and molecular studies (e.g. Janssen et al., 2017) and life cycles suggest the origin of parthenogenesis independently at multiple numbers of time.

Eutelism

In nematodes, cell division during late embryonic development ceases in all organs except the reproductive system and perhaps the midgut and epidermis. For example, the number of cells in the midgut of *Camallanus sweeti* increases from 35 in 1st juvenile (J1) to 200 in the adult (Moorthy, 1938). Rusin and Malakhov (1998) also reported that hypodermal cells are not eutelic in six free-living nematode species. In all other tissue types, the number of cells or nuclei remains constant after hatching. The number of these cells ranges from 434 in *Turbatrix oceti* (Pai, 1927) to 619 *Caenorhabditis elegans* (Bird and Bird, 1991). The distribution of these eutelic cells are indicated below:

Cells (no.)	*T. oceti*	*Rhabditis* ♀	*C. elegans**
Total	-	560 + 120**	959
Eutelic	434	440	619
Nerve	251 (57.8%)	200 (45.5%)	-
Digestive tract	102 (22.5%)	172 (39.1%)	-
Muscle	64 (14.7%)	68 (5.5%)	-
Mesenchyme	16 (3.6%)	-	-
Excretory	1	-	-

* = Sommer, 2015; ** = epidermal cells

According to Bird and Bird (1991), the final number of cells is 959 in hermaphrodite and 1031 in male *C. elegans*. Nevertheless, the nematode body consists of 434 to 619 eutelic cells. Of them, the proportion decreases in the following descending order: Nerve cells > digestive tract cells > muscle cells > mesenchymal cells. Hence, the nervous and digestive systems dominate the eutelic anatomy of nematodes. It is notable that the nematodes are also characterized by less than a dozen cell types, which seem to have facilitated the repeated emergence and evolution of parasitic species.

In pseudocoelomates, eutelism seems to be associated with high temperature tolerance (e.g. Rotifera: 46°C by *Philodina rosella*), anaerobiosis (e.g. *Panagrellus redivivus*) and abundance (e.g. Gastrotricha: 1.8 million/m², soil nematodes: 1 million/m²). In them, tolerance to high temperatures in hot water springs is not uncommon, for example, 52°C to 61.3°C

TABLE 12.3

Incidence of nematodes in hot-water springs and ability to revive after periods of drying and freezing

Species	Temperature (°C)	Location	Reference
Aphelenchoides sp	61.3	New Zealand	Rahm (1937)
Aphelenchus sp	57.6	Chile	Rahm (1937)
Plectus sp	57.6	Chile	Rahm (1937)
Monhystera filiformis	53.5	Iceland	Hyman (1951b)
Darylaimus aratus	53.0	USA	Hoeppli (1926)
Euchromadora striata	52.0	Italy	Meyl et al. (1954)
Monhystera ocellata	52.0	Italy	Meyl et al. (1954)
Monhystera gerlachii	52.0	Italy	Meyl et al. (1954)
Theristus pertenuis	52.0	Italy	Meyl et al. (1954)
Ability to revive after desiccation/freezing			
Species	**Location**	**Enduring Period (y)**	**Reference**
Anguina tritici	Galls	9–27	see Hyman (1951b)
Ditylenchus dipsaci	Galls	4–9	
Platylenchus rotensis	Fig roots	11	
Tylenchus polyhypnis	Rye leaves	39	
Plectus, Dorylaimus	Freezing in moss	2	

(Table 12.3). However, the number of such nematode species and their worldwide distribution is remarkable. Interestingly, the ability of some nematodes to endure desiccation and freezing is amazing. The tylenchids are notoriously resistant to desiccation. Some of their juveiles can be revived from dry galls, roots and leaves after 4 to 27 years. The adult of many species can also endure long periods of desiccation as well as hard freezing. A number of reports are available on revival of live nematodes after wetting or thawing from the dried or hard frozen mosses and soils. Spitsbergan mosses kept dry for 2 years yielded live specimens of *Plectus*, *Dorylaimus* and *Cephalobus*. Live nematodes were revived on wetting samples of moss kept dried in a herbarium for 4.5 years (Hyman, 1951b).

Anaerobiosis

Chemical composition of free-living nematodes ranges from 2 to 19% for glycogen and 23 to 36% for lipids. Conversely, the range for intestinal parsasitic nematodes is 12 to 55% for glycogen and 3.5 to 9% for lipids.

Free-living nematodes utilize glycogen under anaerobic conditions and lipids, when oxygen is available (Schiemer 1987). The intestinal parasitic nematodes, which are subjected to almost continuous exposure to anoxia, rely totally on glycogen. Subjected to anaerobiosis in anoxic sediments, the free-living nematodes, for example, *Panagrellus redivivus* produce a range of reduced organic end products; of them, the major (85%, i.e. 500 of 589 nmol/mg dry tissue) neutral end products are ethanol and glycerol (Table 12.4). The acidic end products, which constitute 13% (~ 78 of 589 nmol/mg), include alanine, lactate, propionate and acetate. Lactate is more rapidly excreted than the others. The nematode is capable of remetabolizing the accumulated end products. The formation of ethanol and glycerol as anaerobic end products is extremely unusual. Ethanol production is more characteristic of prokaryotes than eukaryotes (Muller et al., 2012). The most likely pathway of ethanol production is decarboxylation of pyruvate to acetyledehydrate followed by reduction to ethanol. The decarboxylation of pyruvate is a source of anaerobically produced CO_2 at the rate of 2.3 ml/mg dry weight of *P. redivivus*. Metabolization of glycogen to ethanol and glycerol involves energetically less efficient pathways (see Pandian, 1975). For example, glycerol formation uses ATP rather than producing it. Unlike lactate production, the glycerol production is not in redox balance; it is an electron sink. However, it is an adaptation to cold and desiccation. Similarly, ethanol production is also advantageous, as it prevents tissue acidification and may serve as a mechanism for reoxidizing mitochondrial NADH (Butterworth and Barret, 1985).

In vertebrate parasitic nematodes, a high CO_2 level in the gut leads to tissue acidification. They excrete most acids quickly, as they are removed by the hosts and thereby prevent excessive local pH changes (Butterworth and Barret, 1985). In *Ascaris lumbricoides*, for example, glucose is converted into

TABLE 12.4

Major organic end products during aerobic and anaerobic metabolism in the free-living nematode *Panagrellus redivivus* (condensed from Butterworth and Barrett, 1985)

End Product	Aerobic Tissue Level	Anaerobic Tissue Level	Excretion Rate (nmol/mg dry weight/d)
	(nmol/mg dry weight)		
Ethanol	4.5	312	614
Glycerol	19	188	214
Glycerol-3-phosphate	4.5	11.3	-
Alanine	33	52.4	40
Lactate	8.1	14.8	68.6
Propionate	1.2	6.3	22.5
Acetate	0.8	4.6	17

succinate and alanine. Succinate may either accumulate or be converted into propionate. The conversion of succinate to propionate is energetically more advantageous, as it generates 2 M of ATP (see Pandian, 1975).

Free Living Nematodes

Because of their economic importance, more information is available for parasitic than for free-living nematodes. Poulin and Morand (2000) estimated that > 10,500 nematode species are parasites. Considering nematodes to consist of 26,646 species, ~ 3,500 and 8,360 nematode species were identified as parasites of invertebrate and vertebrate hosts, respectively (Hugot et al., 2001). Recently, Bernard et al. (2017) identified 4,100 species as plant parasitic nematodes. In a broad estimate, Anderson (2000) suggested that ~ 33% of all the described nematode genera occur as parasites on vertebrates, which is equal to the percentage of genera known in marine and freshwater. Hence, the number of nematodes is assessed at ~ 4,100, 3,500 and 8,360 species as parasitic on plants, invertebrates and vertebrates, respectively (however, see also Morand et al., 2005). On the whole, of 27,333 nematode species, 15, 13 and 31% are parasitic on plants, invertebrates and vertebrates. This may indicate that ~ 41% nematode species are free-living. As a large number of nematodes remains to be identified and described, the number may change but not the proportion. Despite constituting 41%, the free-living nematodes have not received much attention, in comparison to parasitic nematodes. While some are herbivores, saprovores or predators, a vast majority of them are bacterivores. The feeding rate of nematodes is far higher for bacteria than for fungi or diatoms (Schiemer, 1987). However, densities higher than 1×10^6 cell/ml are required to grow and reproduce (e.g. *Caenorhabditis elegans*, Table 12.5).

Based on the Life Span (LS), Generation Time (GT) and fecundity, Schiemer (1987) recognized three basic types, of which relatively more information is available for the first two types: Type 1 is characterized by short GT and longer Reproductive Life Span (RLS) and attains a reproductive peak immediately after maturation with Lifetime Fecundity (LF) of 50–500 eggs (e.g. *Caenorhabditis elegans*, Fig. 12.3A). Type 2 includes species with long LS and GT as well as gradual increase in fecundity following maturation and LF of 100–1000 eggs (e.g. *Plectus palustris*, Fig. 12.3B, Schiemer et al., 1980). In them, bacterial (food) density has a profound effect on growth and possibly LF. Interestingly, the smaller nematodes like *Chromadorina germanica* and *Monhydstera disjuncta* (0.4 µg) allocate 67–71% of its somatic production on reproduction. With increasing body size to 1.5–1.6 µg, *P. palustris* and *Rhabditis marina* allocate 60–88% of the production on reproduction. However, with 2 µg size, *Aphelenchus avenae* is able to allocate only 48%

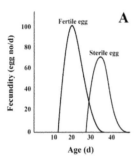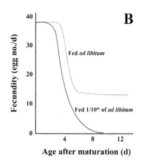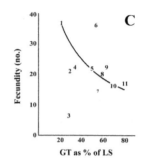

FIGURE 12.3

A. Effect of age on fertility of *Caenorhabditis elegans* eggs (modified and redrawn from Klass, 1987).
B. Effect of ration on fecundity of *Plectus palustris* (modified and drawn from Schiemer, 1987).
C. Fecundity as function of generation time (GT) in a few marine free-living nematodes (type 3). 1 = *Monhystera disjuncta*, 2 = *M. denticulata*, 3 = *Oncholaimus brachycercus*, 4 = *M. denticulata*, 5 = *Desmodera scaldensis*, 6 = *Diplolaimeloides oscheri*, 7 = *M. denticulata*, 8 = *Chromadorita tenuis*, 9 = *D. islandica*, 10 = *M. disjuncta*, 11 = *Adoncolaimus thalassophygas* (drawn from relevant data reported by Zaika and Makarova, 1979).

of its production on reproduction (Schiemer, 1987). Consequently, the LF progressively decreases from 1.3–1.5 times the body weight to 0.8 time of the body weight. Incidentally, the values reported by Zaika and Makarova fall between the LF of 30 eggs for *Theristus pertenuis* and 70–100 eggs for *Rhabditis marina* (see Hope, 1974). Type 3 shows a long GT followed by short RLS and low fecundity (e.g. Zaika and Makarova, 1979). For type 3, data reported by Zaika and Makarova (1979) for 9 nematode species inhabiting marine sediments on Life Span (LS) and Generation Time (GT) were calculated for GT as percentage of LS. On plotting the values, the emerging trend clearly shows that fecundity decreases with increasing GT as % of LS (Fig. 12.3C).

The following two examples provide relevant information on the profound effects of bacterial (food) density and temperature on survival, growth and reproduction in a protandric hermaphrodite *C. elegans* and three gonochoric nematodes. Remarkably, even in the absence of bacterial food, *C. elegans* lived for 4 days, perhaps osmotrophically drawing dissolved organic nutrients from bacteria-free medium (Table 12.5, Chia and Warwick, 1969). However, a minimum density of 1×10^4 cells/ml and 1×10^7 cells/ml are required to sustain growth and reproduction, respectively. Whereas Life Span (LS) is extended from 5 days at 1×10^4 cell/ml to 26 days at 1×10^8 cell/ml, it decreases from 19 at 5×10^8 cell/ml to 15 days at 1×10^{10} cell/ml. At these bacterial densities, Lifetime Fecundity (LF) is increased to 206–273 eggs. Interestingly, the restriction of bacterial density to 1×10^8 cell/ml is extended LS by 52 and 19%, when it is commenced on the first day of growth and reproduction, respectively. However, the restriction on the 1st or 8th day extends LS only marginally by 19 or 12% during menopause. Apparently,

TABLE 12.5

Effect of bacterial density on life span and fecundity of *Caenorhabditis elegans* (modified from Klass, 1977)

Bacterial Density (no./ml)	Life Span (d)	Fecundity (egg no.)
0	4	0
1×10^4	5	0
1×10^6	5	0
1×10^7	15	14
1×10^8	26	63
5×10^8	19	206
1×10^9	16	273
1×10^{10}	15	26

(i) egg production occurs at the cost of life span and (ii) restriction of food density, especially during growth and reproduction extends LS.

In animals including nematodes, LS includes (i) pre-reprodutive duration or generation time, (ii) Reproductive Life Span (RLS) and (iii) post-reproductive or menopause duration. Klass (1977) also estimated the effect of temperature on these durations as well as fecundity in *C. elegans* fed on excessive bacterial food. At 6°C, *C. elegans* did not reproduce. Between 10°C and 25.5°C, temperature increase reduced LS from 35 days at 10°C to 9 days at 25.5°C. But LF increased from 84 eggs at 10°C to ~ 270 eggs at 20–24°C. Apparently, higher temperatures increase egg production but at the cost of LS (Table 12.6). Clearly, factors like food and temperature increase fecundity at the cost of LS.

TABLE 12.6

Effect of temperature on life span and fecundity of *Caenorhabditis elegans*. LS = life span, GT = generation time (number of days in pre-reproductive phase ÷ LS), RLS = reproductive life span (number of reproductive days ÷ LS), Meno = menopause (number of post-reproductive days ÷ LS), Fecund = fecundity in number of fertilized eggs, Hatch = hatchability, Unfert = number of unfertilized eggs (recalculated data from Klass, 1977)

Temperature (°C)	LS (d)	GT (%)	RLS (%)	Meno (%)	Fecund	Hatch (%)	Unfert
6	17.8	-	-	-	-	-	-
10	34.7	28.8	40.3	30.8	84	58	1
14	20.8	24.0	36.0	30.0	206	92	45
16	23.0	17.4	30.4	51.2	250	93	43
20	14.5	20.6	41.4	38.0	273	95	125
24	9.9	20.2	40.4	39.4	269	99	135
25.5	8.9	22.5	44.9	32.6	103	93	51

In a more comprehensive study, Woombs and Laybourn-Parry (1984) investigated survival, growth and reproduction in gonochoric nematodes

Diplogasteritus nudicapitatus and *Paroigolaimella bernensis* (Diplogasterinae), and *Rhabditis curvicaudata* (Rhabidinae) under conditions of excess agar food within a temperature range of 5°C–20°C. (1) In all three species and at all tested temperatures, survival or longevity decreased in the following order: virgin female > male > non-virgin female (Fig. 12.4A, B, C). Only at 5°C, *D. nudicapitatus* male outlived the virgin. (2a) In all three species, growth rate was accelerated with increasing temperature. Adult size varied significantly with temperature in *D. nudicapitatus* and *R. curvicaudata*. (2b) In all three species and at all tested temperatures, males grew to ~ 50% of the maximum size of the respective females (Fig. 12.4D, E, the trend for the diplogasterin *D. nudicapitata* is similar to the other diplogasterin *P. bernensis*; hence, not shown). (3a) Notably, reproductive life span (RLS) was altered in all three species; for example, RLS commenced on the 5th day after hatching and ceased on the 15th day in *R. curvicaudata* at 20°C. But it lasted from the 27th day to 38th day at 10°C. (3b) *P. bernensis* did not reproduce at 5°C but showed a progressive increase in batch fecundity, as in the others (Fig. 12.4F). (3c)

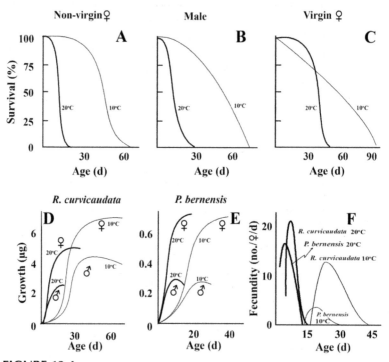

FIGURE 12.4

Survival of A. non-virgin female, B. male and C. virgin female of *Rhabditis curvicaudata* reared at 10°C (thin line) and 20°C (thick line). D. Sex-dependent growth of *R. curvicaudata* and E. *Paroigolaimella bernensis* and males. F. Fecundity of *R. curvicaudata* and *P. bernensis* reared at 10°C and 20°C (compiled, simplified and redrawn from Woombs and Laybourn-Parry, 1984).

LF also increased with increasing temperature; for example, LF increased from 32 eggs at 10°C to 243 eggs at 20°C in *D. nudicapitatus*; these values are 35 and 107 eggs for *P. bernensis*. But it decreased beyond 15°C in *R. curvicaudata*. Clearly, egg production occurs at the cost of LS in both gonochoric and hermaphroditic (Klass, 1977) nematodes. (4) Woombs and Laybourn-Parry noted that 6.5% of mature females failed to produce eggs at all temperatures, but they did not report the number of fertile and infertile eggs as well as the duration of menopause. (5) All free-living nematodes undergo a short or longer period of aging/sterility/menopause.

Parasitic Nematodes

Because of their economic importance, parasitic nematodes have attracted a large number of publications. Molecular analyses have shown that nematode parasitism on plants and animals have originated independently at least 15 times (however, 7 times, see Sommer, 2015). All parasitic nematodes may have a terrestrial or limnic origin (see Blaxter and Koutsovoulos, 2015). Most terrestrial plants and large animals are associated with at least one species of nematode. On the global scale, the damage cost of parasitic nematodes ranges from 80 to 118–125 billion US$/y, Chitwood, 2003, Table 12.7. The incurred loss increases from 10% in maize to ~ 35% in tomato. Increase in water content of the agricultural product increases the damage level. Parasitic nematodes also inflict a huge loss to humans and their livestock and fish. Though comprehensive and reliable estimates are not available, the damage cost to human health, and livestock and fish may easily account for another 100 billion US$/y. Incidentally, the parasitic trematodes (~ 16,000 species) and cestodes (~ 4,000 species) also inflict equally heavy losses including death of man and his livestock (Pandian, 2020). However, they do not harm his agricultural crops. Hence, the parasitic nematodes inflict dimensionally a greater damage to man and his food basket.

Plant Parasites

All nematodes undergo four molts and have six stages during their development namely the egg, 4 juvenile and adult stages. In animal parasitic nematodes, mostly juvenile stage 3 (J3) is infective but the migratory J2 is infective in plant parasitic nematodes; the plant infection is always by penetration. The root-knot nematodes (e.g. *Meloidogyne* spp) target the undifferentiated procambium cells of the vascular cylinder. Their life cycle is completed, when their eggs are released into soil. Their infection causes stunted growth, wilting and root galling. In contrast, the cyst nematodes (e.g. *Heterodera* spp) enter the root tips and induce syncytia by destroying cell

TABLE 12.7

Economic loss caused by parasitic nematodes to man and his crops, livestocks and fish. GI = Gastrointestinal, bil = billion, mil = million, mmt = million metric ton

	Comment	Reference
Human		
GI nematodes	3.5 bil affected, 4.5 mil seriously ill, 125,000 death/y	Holland (2013)
Cattle		
GI nematodes	0.5 kg–1.4 kg meat reduction/cow/d	Rodriguez-Vivas et al. (2017)
Ostertagio ostertagi	1.25 l milk reduction/cow/d	Ravinet et al. (2016)
Sheep		
GI nematodes	Loss is 15% for meat, 10% for wool and 22% for milk production	Mavrot et al. (2015)
Goat		
GI nematodes	Loss of $ 5.5/goat/y	Shakya et al. (2017)
Agricultural crops (from Bernard et al., 2017 & others)		
Heterodera avenae, H. filipjevi, H. latipons, Hordium vulgare	From global production of 758 mmt wheat *Triticum aestivum*, 15% , i.e. 83 mmt loss/y due to cyst nematode infection. 30–100% loss is incurred in different countries	
Meloidogyne graminicola, Hirschmanniella oryzae; other 100 nematode species	From global production of 480 mmt rice *Oryza sativa*, 15%, i.e. 53 mmt loss/y due to rice root nematodes. *H. oryzae* can cause up to 80% loss	
Meloidogyne spp, *Pratylenchus* spp, *Heterodera* spp, ~ 50 species	10.2% loss of maize *Zea mays* due to root-knot, root lesion and cyst nematodes	
Globodera pallida, G. rostochiensis, Ditylenchus destructor	From global production of 376 mmt potato *Solanum tuberosum*, 15% , i.e. 41 mmt due to root-knot, stem nematodes. *Ditylenchus destructor* can cause up to 100% loss	
Root-knot nematodes and others	From global production of 105 sweet potato *Ipomoea batatus*, 10.2% , i.e. 10 mmt due to root-knot nematodes	
Meloidogyne incognita	28.6% loss of cowpea *Vigna unguiculata* (Kar et al., 2018)	
M. artiella, M. incognita	14% loss of chickpea *Cicer arietinuem* (Zwart et al., 2019)	
Root-knot nematodes	9% loss of soybean *Glycine max* (Fourie et al., 2017), i.e. US $/y 1.6 billion in US alone (Li et al., 2010)	
Pratylenchus penetrans, P. neglectus	31% onion loss due to root lesion nematode (Hafez and Palanisamy, 2016)	
M. incognita	~ 35% loss of tomato (Hema and Khanna, 2018)	

walls. Their eggs remain within the roots and are protected until hatching under favorable conditions (Bernard et al., 2017). J2 of root-knot nematodes move intercellularly, whereas J2 of cyst-nematodes pierce and enter the cells one by one and thereby inflict a greater damage to the host plants (Holbein et al., 2016). The lesion nematodes feed on the root cortex and enter the vascular tissues to obtain nutrients. They cause lesion and necrosis (Bernard et al., 2017).

According to Holbein et al. (2016), plants can recognize the nematode infection. The first level of defense is to reinforce the cell wall. Substances like suberin, lignin and/or callose are deposited between cell wall and plasma membrane. For example, the deposition of callose leads to a strong barrier for piercing by the nematode stylet. The second line of defense is to secrete arsenals that include a broad diversity of constitutively produced toxic phytochemicals. A great deal of effort has been made to develop nematode-resistant varieties. At present, development of resistant varieties, crop rotation and/or application of nematicides are the only options to control plant-parasitic nematodes. Resistant crops have a narrower genetic base than that of nematodes and lead to frequent infection by others and crop loss. Crop rotation requires a prolonged interval without the host crop to be effective. Nematicides, including organophosphate and carbamate compounds, are extremely toxic and increase production cost (Li et al., 2010).

With a limited option for an effective and environmentally safe method to prevent or treat nematode infection, molecular biologists have resorted to target the nematode reproduction namely the fitness genes *cpn-1*, *Y25* and *Prp-17*. Introduction of double stranded (ds) RNA depletes the endogenous mRNA in the transgenic plants and thereby reduces egg production by plant parasitic nematodes; the reduction can be 95, 81 and 79%/g root (Li et al., 2010). To introduce the ds RNA, a few methods like microinjection, soaking of the eggs and octopamine-supplemented introduction have been developed. However, they are not feasible on a large scale in fields. Yadav et al. (2006) developed an ingenious method of delivering ds RNA, which ensures total depletion of the target mRNA in *Meloidogyne incognita*. For this, they amplified the $_c$DNAs of two important genes—one coding for the splicing factor and the other coding for integrase from *M. incognita* and cloned them in a plant expression vector that ultimately produced a hair pin-shaped ds RNA. These constructs were then introduced into a tobacco plant with the help of *Agrobacterium*. On inoculation of *M. incognita* J3 into the transgenic plant carrying the constructs, 92% of the transgenic plants were free from *M. incognita*.

In corn nematode *Heterodera zea*, temperature profoundly affects the development and fecundity. Following uniform inoculation density, the J2 penetrated into corn seedling roots and matured within 11–13 days at 33°C–36°C but requires 24 and 38 days at 25°C and 20°C, respectively. However, the number of egg/cyst is less (190 eggs/cyst) at 33°C–36°C than that (235 egg/cyst) at 25°C. Hence, the negative effects of low temperature on development and delayed egg production seem to be compensated by more number of egg/cyst (Hutzell and Kursberg, 1990).

Invertebrate Parasites

Approximately, 3,827 mermithid species are parasitic on invertebrate host like insects, spiders, crustaceans and snails. Whereas mermithid adults lead free-existence, their juveniles are parasitic. They enter the host by ingestion of eggs (e.g. blackflies) or through penetration. Provided with dorylain stylet, most juvenile parasites penetrate the host aided by the propelling modification of the posterior part of the body (Hyman, 1951b). On gaining entry into the host, some mermithids release symbiotic bacteria. For example, *Steinernema* spp carry *Xenorhabdus* spp, *Heterorhabditis* spp and *Rhotorhabdus* spp (Sommer, 2015). These bacteria kill the insect by secreting a number of toxins and thereby provide the carcass with the saprozoic mermithids. While these mermithids are saprophagous feeding on the carcass, the others, on attaining adulthood, leave the hosts after feeding the hemocoel and fat bodies. There seems to be a succinct difference between those mermithids that feed on the carcass and complete a larger fraction of their Life Span (LS) and those that feed on hemocoel and complete a short duration of LS. For example, the share of carcass-feeding juvenile stage in *Agamermis decaudata* is 92.4% of its LS (see Table 12.12). However, it is only ~ 30% for those, for whom mosquitoes serve as host (see Nematology Laboratory, University of Nebraska-Lincoln – website). Table 12.8 lists the number of aquatic insect species and limited publications on nematode parasites on some of them. Being terrestrial, the orthopterans, in which the majority of mermithids are parasitic, are not included in the table. Not surprisingly, very limited information is available for mermithids. Among aquatic insects, mosquitoes (e.g. Dong et al., 2014), chironomids (e.g. Johnson, 1955), simulid blackflies (e.g. Poinar et al., 1976) and trichopteran caddisflies (Lancaster and Bovill, 2017) are parasitized by mermithids. Seven mermithid genera are parasites of mosquitoes: *Culicimermis, Empidomermis, Octomyomermis, Perutilomermis, Romanomermis* and *Strelkovmermis* (Kobayashi et al., 2012). To his credit, Petersen has a series of publications, which explore the limited scope for mosquito control through mermithids. Petersen (1975) is perhaps the only publication, which reports that while active swimming lasts up to 120 days in *Reesimermis nielseni*, its infectivity ceases on the 72nd day.

In *Raphidascaris acus*, a fish parasite, the gravid females size ranges from 0.7 mg to 61.2 mg, a 90-fold difference between the smallest and largest worm (Szalai and Dick, 1989). Larger females produce more eggs per unit time than their smaller conspecific (e.g. Sinniah and Subramaniam, 1991).

TABLE 12.8

Aquatic insect species number and relevant publication number on parasitic nematodes (condensed from Kohler, 2008)

Taxon	Species (no.)		Publication (no.)
Ephemeroptera	3,100		15
Plecoptera	2,000		-
Odonata	5,500		1
Hemiptera	2,600		
Trichoptera	11,000		2
Megaloptera	300		1
Coleoptera	9,500		2
Diptera	30,600		
Ceratopogonidae		5,300	6
Chironomidae		5,000	22
Culicidae		3,500	33
Simuliidae		1,800	33
Others		> 15,000	3
Total	64,600		118

In this context, Poulin and Lotham (2002) examined inequalities in size and intensity-dependent growth of a mermithid parasite in beach hopper crustacean *Talorchestia quoyana*. A consequence of increasing parasite density from one to 199/host is the emergence of a large α-worm, whose size is increased from ~ 82 mm length in a hopper with 6–9 worms to ~ 115 mm in the hoppers harboring one worm/host. The main cause for the inequality in body size of the parasitic worm is the competition for the space and resource.

On the other side, Laws (2009) explored how the host density affects reproduction in grasshopper *Melanoplus dawsoni*, a small hopper with five nymphal instars lasting for 21–66 days. Its field density ranged from 6.7 to 8.9 hopper/m². Correspondingly, the mermithid prevalence also increased from 15 to 37.5% with an average of 27.5%. Experimental alterations of hopper density to 0.25, 0.5, 1.0, 1.25 and 1.5 times, considering field density of 8.9 hopper/m² as 1.0, resulted in reductions in the number of ovarian follicle (potential fecundity) and egg (realized fecundity)/♀ as functions of hopper density and parasitism. Potential fecundity decreased from ~ 10.05 follicle/♀ at 0.25 hopper/m² to ~ 8.5 follicle/♀ at 1.0 hopper/m² (Fig. 12.5A). Correspondingly, the realized fecundity also decreased from ~ 7.5 egg/♀ to 5.5 egg/♀, i.e. the reductions were 20% for potential fecundity but 27% for realized fecundity. Increase in hopper density reduced the resource

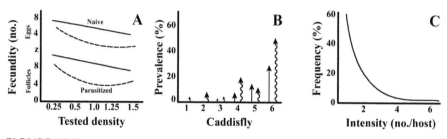

FIGURE 12.5

A. Effect of host density and mermithid parasitism on potential (follicle, lower panel), and realized (egg, upper panel) fecundity in grasshopper *Melanoplus dawsoni* (modified and compiled from Laws, 2009). B. Prevalence of mermithids on male (↑) and female (↑) caddisflies. 1 = *Ecnomus cygnitus*, 2 = *E. turgidus*, 3 = *E. continentalis*, 4 = *E. pansus*, 5 = *E. tillyardi*, 6 = *E. russellius*. C. Effect of mermithid intensity on frequency of caddisflies (modified from Lancaster and Bovill, 2017).

availability to realize potential fecundity. At the mermithid-infected hoppers, 0.25 hopper/m² density level, the reduction at the low density was from 11 follicle/♀ to 8.25 follicle/♀, while it was from 5.75 egg/♀ to 1.5 egg/♀ at the hopper density of 1.0/m², i.e. the reduction was 25% for potential fecundity but as high as 74% for realized fecundity. A combination of hopper density and mermithid parasitism reduced potential fecundity from 10.25 follicle/♀ to 8.5 follicle/♀ and realized fecundity from 7.5 egg/♀ to 1.5 egg/♀. Briefly, increase in density and parasitism reduced potential fecundity marginally by 20% but the realized fecundity by 82%. Hence, realized fecundity is more severely affected by parasitism than potential fecundity.

Examining 40,000 specimens over a period of 4 years, Lancaster and Bovill (2017) estimated the prevalence and intensity of multiple mermithids on six congeneric host caddisflies of *Ecnomus*. In them, prevalence of mermithids ranged from 1 to 4% in *E. continentalis* and from 30 to 50% in *E. russellius*. Male caddisflies are preferred over females by the mermithids. Figure 12.2B shows the total or nearly total absence of mermithids in *E. cygnitus* and *E. turgidus*, respectively. It was less for females in others too. Intensity was one per host in 60% of these caddisflies (Fig. 12.5C). In others, it ranged from two to six per host. The scope for sexual reproduction in 60% of the flies is less. It is not known whether they reproduce, if so, whether they do it by parthenogenesis or by selfing, as hermaphrodites.

Vertebrate Parasites

A glance over Hyman (1951b) reveals that nematode parasites have infested almost all animal taxa and almost all organs, except perhaps the brain. Barring the saprophagous mermithids and *Capillaria hepatica*, no vertebrate parasite is known to kill its host. Its strategy is to sustain parasitism by keeping the host alive but not active and healthy. In vertebrate hosts, the routes of nematode infection include: (1) Trophic or oral ingestion, (2) Skin penetration by J3 (e.g. Ancylostomidae: *Ancylostoma duodenale, A. canicum, Necator americanus*; Strongyloididae: *Strongyloides stercoralis*; Trichostrongyloidae: *Nippostrongylus brasiliensis*, Koprivnikar and Randhawa, 2013) and (3) Injection by infective J3 by blood sucking arthropod Intermediate Host (IH) (e.g. Filarioidea: *Wuchereria bancrofti* by mosquitoes, *Acanthocheilonema grassi* by tick *Rhipicephalus, Dipetalonemaarbuta* by *Aedes, Setaria cervi* by stablefly *Stomoxys* and *Mansonella ozzardi* by sand midge *Culicoidesfurens*, Hyman, 1951b). A vast majority of nematodes infect the Definitive Host (DH) and intermediate host (IH) by oral ingestion. They can further be divided into (a) ingestion of uncleaved eggs by coprophagous rodents (e.g. Oxyuridae), (b) ingestion of cleaved eggs at earlier than J3 eggs (e.g. Ascaridae: *Ascaris lumbricoides* infective eggs containing J2 stage; Oxyuridae: *Oxyuris equi, Enterobious vermicularis*, Hyman, 1951b), (c) ingestion of eggs containing J3 by intermediate host (e.g. Syngamidae: *Syngamus, Stephanurus*, Mehlorn, 2016; Spiruridae: *Spirocercalupi, Seurocyrnea coloni*, Hyman, 1951b) and (d) ingestion of eggs containing encysted J3 by intermediate host (e.g. Trichinellidae: *Trichinella spiralis*; Cystoopsidae: *Cystoopis acipenseri*, encysted in the amphipod host, Hyman, 1951b). Notably, the infective eggs containing J3 stage of *Syngamus* and *Stephanurus* may either directly be ingested by DH pig and fowl, respectively or they may pass through the facultative IH earthworm (Mehlorn, 2016).

Rhabditoids and oxyuroids include both free-living and parasitic nematodes. The other eight nematode orders include exclusively vertebrate parasitic species (Table 12.1). Their direct life cycle involves no IH or may involve one or two IH(s). Some of these IHs are designated as Transport Host (THs), which may not obligately required for completion of the life cycle. For example, the infective eggs of *Stephanurus* and *Syngamus* may directly be ingested or through TH namely the infected earthworms. However, the eggs of strongylids (e.g. *Metastrongylus elongatus*) and spiruroids (e.g. *Spirurua talpae*) do not develop until eaten by the IH namely earthworm/snail/slug/fly/centipede and cricket/grassoppher, respectively. These obligatory hosts may be designated as IH. In free-living copraphagous *Rhabditis coarctata*, the J3 cannot develop further unless transported to fresh dung by the dung beetle, which may have to be designated as IH and not as TH. To most nematode parasites, one or other invertebrate serves as TH or IH. But transmission of *Spirurura lupi*, parasitic on carnivorous vertebrates, the fowl, reptiles and birds may serve as TH. The spiruruid *Camallanus* inhabiting the digestive tract of aquatic vertebrates requires two IH namely *Cyclops/Notodiaptomus* → small fish → DH. Those IH, in which the development from egg to J3 occurs, suffer reduced survival and fecundity. For example, the terrestrial snail *Cernuella virgata* serves as IH to the rhabditid *Phasmorhabditis hermaphrodita*.

On exposure to R954 (*P. hermaphrodita*) at the density of 50, 100 and 150 J3/m^2, all the snails ceased oviposition between 16th and 24th day and all of them were dead between 28th and 30th day after infection, when the nematode began to breed (Charwat et al., 2000).

Most monogeneans are strictly host specific and are specialists. In digeneans and cestodes, the specificity is progressively diluted not only with regard to DH(s) but also to their intermediate hosts (Pandian, 2020). Interestingly, nematodes are also generalists and not host specific. For example, *Ascaris lumbricoides* occurs in man, apes, pigs, cattle, sheep and squirrels. The anisakid ascaroid *Contracaecum* can engage medusa, *Sagitta*, copepods, amphipods and cephalopods as the Intermediate Host (IH). The plant parasitic nematodes are more generalists than those of animal parasites. The rhabditid *Protylenchus similis* is a root parasite of coffee, tea, banana, pepper, sweet potato, pineapple and so on. A climax is *Heterodera marioni*, a root parasite of 380 plant species (Hyman, 1951b). Interestingly, 50 and 100 nematode species can infect the maize *Zea mays* and the rice plant *Oryza sativa*, respectively (see Bernard et al., 2017). Hence diagnosis, treatment and control programs encounter the greatest challenge.

In the context of specificity and immunity in vertebrate hosts, Koprivnikar and Randhawa (2013) found that some nematodes are specific to a host species. Host specificity is recognized as a key factor that determines distribution of parasitic species. As a 'compatibility filter', it narrows the range of potential hosts, in which the parasite cannot survive and develop. The host immune response may reduce parasitic growth and reproduction. For example, *Strongyloides ratti* female grows to smaller size (~ 1,300 μm in length) and is less fecund (125 egg/♀/d) in the immunized rat than in the naïve (non-immunized) rat growing to a length of 1,800 μm and ovipositing 180 egg/♀/d (Viney et al., 2006). Similarly, the protective vaccination of a dog with *AC-16*, an immunodominant surface antigen from the hookworm reduced the fecundity of *Ancylostoma canicum* to 36% (Fujiwara et al., 2007). Koprivnikar and Randhawa (2013) listed information on fecundity for 19 and 5 nematode species, in which the routes of infection are by ingestion and penetration, respectively. The fecundity ranged from 50 to 17,500 egg/♀/d for penetrating nematodes but 17 to 270,000 egg/♀/d for the trophic transmitters. However, their mean values are 37,348 egg/♀/d and 7,195 egg/♀/d for the trophic and penetrative nematodes, respectively. The penetrative mode of infection seems to demand relatively less resource for egg production and consequently, they may be less fecund.

Prevalence and Intensity

Prevalence and intensity of nematode infection may depend on a number of factors such as fecundity, inclusion of IH and their motility, as well as host susceptibility and density. Available (though not exhaustive) information on prevalence and intensity of parasitic nematodes involving the inclusion of no (Table 12.9) and one (e.g. *Rhabdias bufonis*) or two (e.g. *Camallanus* spp, Table 12.10) IH is listed. The listed information suggests that (i) prevalence is higher and ranges from 71–88 to 99.9% in nematodes involving one or two IHs. In the nematodes involving no IH, however, the level is relatively lower and ranges from 1.5% in *Capillaria caudinflata* to 66% in *Capillaria catostomi*

TABLE 12.9

Prevalence and intensity of nematode parasites involving no intermediate host

Order/Family/Species	Prevalence (%)	Intensity (no./host)	Reference
Strongyloidea			
Strongylidae			
Strongyloides stercoralis	30.4		Muazu et al. (2017)
Strongyloides sp	12.0		Chavhan et al. (2008)
Strongyloides sp	4.4		Ayeh-Kumi et al. (2009)
Metastrongylidae			
Metastrongylus (5 species)	97		Humbert and Henry (1989)
Ascarops stronglina	92		Humbert and Henry (1989)
Ancylostomatidae			
Bunostomum on cow	1.4		Gunathilaka et al. (2018)
Ancylostoma duodenale	2.0		Ayeh-Kumi et al. (2009)
Trichostrongyloidae			
Haemonchus	3.8		Chavhan et al. (2008)
Ascaroidea			
Ascaridae			
Ascaris lumbricoides	84–87	11	Hall and Holland (2000)
Ascaridia galli	63.8		Permin et al. (1999)
Toxocara	27.7		Chavhan et al. (2008)
T. canis	14.9	7.0	Dybing et al. (2013)
Toxascaris leonina	4.7	1.0	Dybing et al. (2013)
Oxyuroidea			
Oxyuridae			
Capillaria catostomi on fish	66.0	9.6	Bell and Bereley-Burton
	53.6		(1980)
C. obsignata	32.0		Permin et al. (1999)
C. anatis	4.1		Permin et al. (1999)
Enterobius vermicularis	1.5		Ayeh-Kumi et al. (2009)
C. caudinflata			Permin et al. (1999)
Spiruroidea			
Spiruridae			
Oxyspirura petrowi	32.0	9.6	Kubecka (2017)
Trichuroidea			
Trichuridae			
Trichuris	14.9		Chavhan et al. (2008)
Trichuris on cow	4.8		Gunathilaka et al. (2018)
Trichuris trichura	3.6		Muazu et al. (2017)

TABLE 12.10

Prevalence and intensity of nematode parasites involving intermediate host(s)

Order/Family/Species	Prevalence (%)	Intensity (no./host)	Reference
Rhabditoidea Rhabditidae 　*Uncinaria stenocephala*	18.2	17	Dybing et al. (2013)
Rhabdiasoidea 　*Rhabdias bufonis*	37.1	0.5	Ibrahim (2008)
Ascaroidea Anisakidae in cod 　*Anisakis simplex* in herring	92 99.9		Szostakowska et al. (2005) Szostakowska et al. (2005)
Spiruroidea Spiruridae 　*Spiroxys contortus* Acuariidae 　*Acuria xanthurus* 　*Camallanus maculatus* 　*C. oxycephalus*	 82 71	 2.8	*Cyclops* spp, I IH, *Ameiurus nebulosus*, tadpoles, II IH Hedrick (1935) Martins et al. (2007b) (see Stromberg and Crites, 1974)
Dracunculoidea *Drancunculus medinensis* 　*D. insignis* in dog 　*D. insignis* in human	 2 8		Cairncross et al. (2002) Cairncross et al. (2002)
Filarioidea *Acanthocheilonema reconditum* *Mansonella ozzardi*	12 4–100		Brianti et al. (2012) Shelley (1975), Medeiros et al. (2009)

among oxyuroids. Clearly, the inclusion of one and two IHs increases the prevalence. (ii) However, there are exceptions to this generalization. For example, the prevalence is as high as 84–87% for *Ascaris lumbricoides* involving no IH. This may be traced to the high population density and poor hygienic conditions prevailing in Bangladesh and Nigeria (Hall and Holland, 2000). Likewise, the low prevalence of dracunculoids and filarioids can be traced to penetrative and injective modes of transmission as well as health care and pest control measures in countries, where the surveys were made. (iii) Strikingly, intensity of infection (0.5–17/host) is relatively lower in those involving IH than in those (1.0–11/host) in nematodes involving no IH. It has great implications for sexual or parthenogenic modes of reproduction.

The presence of nematode parasites may allow infection by conspecific but not others. From their studies in the Baltic Sea, Szostakowska et al. (2005) reported an excellent example for it. For *Anisakis*, *Pseudoterranova* and *Contracaecum*, fish-eating mammals and birds serve as DH, and for

Hysterothylacium, smaller and large fish serve as IH and DH, respectively. In raw fish-eating countries, anisakidosis is caused by *A. simplex*, as its prevalence is almost 100%. From the details mentioned below, it may be inferred that the already hosted *A. simplex* in herring and anisakids in cod allow the incoming infection by conspecifics but eliminate others. In *A. simplex*, prevalence in herring increases from 5% at the age of 3 years to 99% of the age of 11 years. This is also true of the cod host. In it, the prevalence of anasakid complex increases from 3.2% in 40 cm size to ~ 58% in sizes of > 75 cm. A similar but not such an intense competitive exclusion of cestode parasite *Bothriocephalus archignathi* by *Camallanus cotti* was reported by Vincent and Font (2003).

Parasite Species	Prevalence (%) in	
	Herring	Cod
A. simplex	99.9	3.73
Contracaeum osculatum	0.015	0.65
Hysterothylacium auctum	0.056	1.46
Rhaphidascaris acus	0.004	0.16
Anisakid complex		92.0

In this context, information available on oxyuroids has relevance. They are known for batch fecundity and periodicity of oviposition. It is not known whether the batching and periodicity are regulated by structural features of reproductive system or limited by sperm availability. Zervos (1988) reported on the monogamic feature in the oxyurid *Blatticola monandros* on the cockroach *Parettipsidion pachycercum*. Only one male and one female *B. monandros* infect 99.85% of the cockroach. The incidence of two females per host is rare. In the presence of a single female, *B. monandros* produce > 4-times as many eggs as produced by two females in a single host. It is not known whether the reduced fecundity in the host with two females is due to competition for resource or chemical interference between the females. With one female in a single host, oviposition is cyclical, i.e. once in 5 days. It is also not known whether it is due to a structural limitation of reproductive system or sperm availability. Zervos (1988) also listed a number of oxyurid species, in which sex ratio is skewed in favor of females and the ratio may have implication to sex differentiation process. For example, *Wetanema hula* hosted by *Homideina thoracica*, *Thelastoma attenuatum* hosted by *Periplaneta americana* and all the thelastomatids hosted by crickets, the ratio is skewed in favor of females. A climax seems to be the ratio of 0.9 ♀ : 0.1 ♂ in *Thelastoma collare* hosted by millipede *Orthoporus ornatus*. In a single host, not more than one male is present for many females of *Leidynema appendiculata* on *P. americana* and *Thelastoma moko* on the millipede *Eumastigonus*. It is not known whether the presence of one male inhibits the other genetic males to differentiate into females.

Likewise, interesting information is also available for the nematodes involving two IH. The camallanids include 360 species and are globally distributed (Stromberg and Crites, 1974). Among them, *Channa gachua*, percids, salmonids and gadids are identified as DH. They are highly specific to the First Intermediate Host (FIH) but the specificity is a little diluted for the Second Intermediate Host (SIH). For example, given the choice of *Cyclops vernalis*, *C. biscuspidatus*, *Diaptomus*, *Asellus*, *Gammarus*, *Hyallela*, *Chironomus*, *Camallanus oxycephalus* selected *Cyclops* only, as 71% infection occurred in *Cyclops* but none in others. Ingestion of *Camallanus oxycephalus*-infected *Cyclops* resulted in 80% prevalence in the yellow perch (Stromberg and Crites, 1974). A minimum of 120 *Camallanus* juvenile/9 ml water at 28°C ensured 46% infection on the copepod *Notodiaptomus* (Martins et al., 2007a). Within poecilids, prevalence of *C. cotti* was 65, 49 and 32% for *Poecilia reticulata*, *P. mexicana* and *Xiphophorus helleri*, respectively. In these fishes, the values for intensity were 2.1/host, 6.1/host and 1.6/host (Vincent and Font, 2003). However, the prevalence of *C. oxycephalus* was dominated by 71% of the gizzard shad *Dorosoma cepedianum* (see Stromberg and Crites, 1974) and 82% for *C. maculatus* on *X. maculatus* (Martins et al., 2007b).

Interestingly, some information is also available on prevalence and intensity of nematode infection in intermediate hosts. Shah and Mohilal (2012) reported about the nematode intensity in soil-dwelling insects of Manipur, India. The intensity ranged from 6.1/host in *Periplaneta americana* to 11.9/host in *Hydrophilus triangularis*. More than the number of earthworm species, its density determines prevalence and intensity of the lung nematodes in a Ukraine piggery farm (Table 12.11). Prevalence decreased from 27% in the presence of 975 earthworms in the soil to 34% in pastures with only 160 earthworms. In contrast, intensity increased from 14/earthworm to 41/earthworm (Antipov et al., 2018).

TABLE 12.11

Prevalence and intensity of lung nematode infection on earthworms in and around an Ukrine piggery farm (condensed from Antipov et al., 2018)

Location	Species (no.)	Density (no.)	Prevalence (%)	Intensity (no/host)
Pigties	2	1160	62.6	22.3
Soil	4	975	26.9	14.0
Pens	4	675	29.9	19.7
Pastures	4	160	33.8	40.7

Fecundity

For nematodes, food consists of (i) bacteria, (ii) plant cells and their sap, (iii) carcass and hemocoel, and (iv) digested food or blood (e.g. hookworms) for the (i) free-living, (ii) plant parasitic, (iii) invertebrate parasitic and (iv) vertebrate parasitic nematodes, respectively. Though bacterial and plant cells may provide micronutrients (that cannot be synthesized by animals), the animal feed is nutritionally richer than the others. Arguably, the nematode fecundity—reported as Lifetime Fecundity (LF, egg no/♀) or daily fecundity (DF, egg no/♀/d)—progressively increases from the free-living to vertebrate parasitic nematodes. Accomplishing an onerous task, Morand (1996) summarized the relevant information on body size, Life Span (LS), Generation Time (GT) and fecundity. Subjecting the data to appropriate statistical analyses, he made a few generalizations on the relations between fecundity on one hand and body size, LS and GT as % of LS on the other. However, his analyses suffer from the following: (1) Data reported for LF and DF are identified but are not separately considered in the analyses. (2) Free-living nematodes undertake no risk to complete a generation in their life cycle. However, the transmission from one DH to another involves an event of risk in parasites with no IH. But in those with one or two IH(s), the number of risky events are doubled or trebled. Parasites are known to increase fecundity to compensate these risky events (e.g. platyhelminthic parasites, Pandian, 2020). Hence, the need to separately consider the fecundity of the free-living and parasitic nematodes involving no IH or IH is obvious. Though Morand identified them as monoxenous and heteroxenous life cycles, he considered all of them together in his analysis. The number of nematode species considered by him is also not balanced with the proportions of free-living and parasitic nematodes.

In the present analysis, the values for LF and DF are separately arranged (Table 12.12) for free-living, mermithid and vertebrate parasites. In it, LF and DF values for these nematodes are also separately considered. In these nematodes, Life Span (LS) ranges from 7 days in *Acrobelpoides* sp to 63 days in *Dilenchus triformis* for free-living species but from 14 days in *Toxcaracanis* to 2,370 days in parasite *Necator americanus* (Table 12.12). Strikingly, parasitism has extended LS more than hundred times. Not surprisingly, available data for Lifetime Fecundity (LF) for parasites are very limited.

The rearranged data are plotted separately for LF and Daily Fecundity (DF) against (i) body size, (ii) LS (Fig. 12.6) and (iii) GT as % of LS (Fig. 12.7A). From these, the following may be inferred: (1) The cumulative relation for body size vs LF shows a clear positive linear increase of LF with a greater size. The trend for the free-living nematodes including a single value for free-living cum plant parasitic nematode is also positive and linear but the values fall at lower levels than those of parasitic nematodes (Fig 12.6 A). The

TABLE 12.12

Life history traits of oviparous nematodes involving no intermediate host (IH) (rearranged from Morand, 1996). DF = daily fecundity (egg no./♀/d), LF = lifetime fecundity (egg no./worm), * Woombs and Laybourn-Parry (1984); † from Hyman (1951b)

Species	Size (mm)	Life Span (LS) (d)	GT (% of LS)	LF/DF
Free-living nematodes				
Caenorhabditis elegans	1	20	12.5	240
Rhabditis curvicaudata*	-	15	20.0	113
Paragolaimella bernensis*	-	13	23.1	107
Diplogasteritus nudicapitatus*	-	22	13.6	247
Dilenchus triformis	1	63	15.8	79
D. dipsaci	1	60	15.0	352
Mesodiplogaster biformis	1.5	8.5	29.4	240
Acrobelpoides sp	0.6	7	85.7	39
Contortylenchus elongatus	5	-	-	7500
Mermithid parasites				
Agamermis decaudata	280	395	92.4	10000
Reesimermis nielseni	15	-	-	3000
Pheromermis villosa	52	-	-	30000
Vertebrate parasites				
Enterobius vermicularis	10	60	36.7	11000
Aspicularis tetraptera	4	-	-	482
Passalurus ambibuus	9	-	-	1500
Sphacia obvelata	6	-	-	1500
DF in vertebrate parasitic nematodes involving no IH				
Parascaris equorum	150	270	34.4	270000
Ascaris lumbricoides	305	548	11.9	103500
Nippostrongylus brasiliensis	7	15	40.0	560
Ostertagia ostertagi	9	365	57.5	200
Trichostrongylus tenuis	9	365	-	356
Necator americanus	10	2370	1.8	15000
Nematospiroides dubia	12	84	11.9	600
Ancylostoma duodenale	14	1640	2.4	23000[†]
Amidoostomum anseris	17	200	-	284900
Haemonchus contortus	24	300	-	5000

TABLE 12.12 Contd. ...

...TABLE 12.12 Contd.

Species	Size (mm)	Life Span (LS) (d)	GT (% of LS)	LF/DF
DF in viviparous nematodes involving IH				
*Pseudoterranova dicipiens**	58	18	-	7000
Acanthocheilonema vittae	70	450	50	188
Dipetalonema viteae		52	-	7000
Litosomoides carini	120	360	20.5	15000
Toxacara cati	70	90	43.3	171000
T. canis	150	14	43.8	280000
Wuchereria bancrofti	100	1640	11.1	10000
LF in viviparous nematodes involving no IH				
*Trichuris muris**	40	39	35.9	6000
T. trichura	60	720	6.8	5000

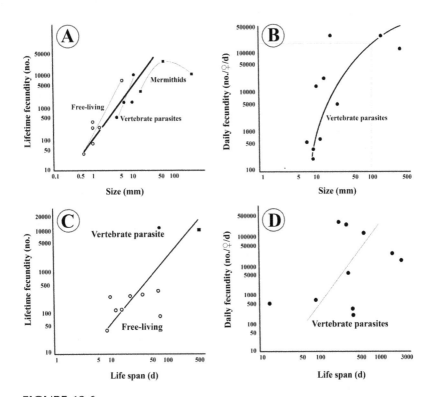

FIGURE 12.6

A. Life time fecundity as a function of body size, B. daily fecundity as a function of body size in nematodes. C. Life time fecundity as a function of life span, D. daily fecundity as a function of life span of nematodes (drawn from data reported in Table 12.11).

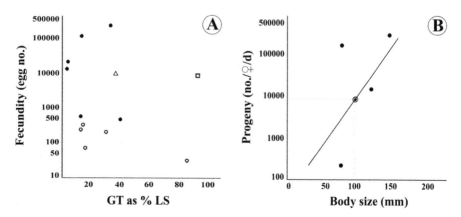

FIGURE 12.7

A. Lifetime fecundity (open symbols) in free-living (○), mermithid (□) and vertebrate parasitic nematodes (△) and daily fecundity (●) in parasitic nematodes as a function of GT as % LS. B. Daily fecundity in viviparous nematodes. Note a single value (◉) in Fig. 12.7B is related to viviparous *Wuchereria bancrofti* transmitted by *Cyclops* (drawn from data reported in Table 12.11).

scattered values of mermithid is positive but more asymptotic than linear. This may be traced to the difference in the share of the juvenile phase as % of LS in carcass-feeding and hemocoel-feeding mermithids. (2) The trend for the size and DF relation is also positive and almost linear for not only oviparous (Fig. 12.6C) but also for viviparous (Fig. 12.7B) parasitic nematodes. However, it may be noted that (i) the body size of viviparous filarioids is limited to 150 mm (Fig. 12.7D) but those of oviparous nematodes is extended up to 500 mm (Fig. 12.6B,C). At a comparative body size, of about 100 mm length, the oviparous parasitic nematodes involving no IH are more fecund (120,000 egg/♀/d, Fig. 12.6B) than the viviparous/filarioids (10,000 progeny/♀/d, Fig. 12.7B), whose transmission involves mostly (injective) hematophagous IH. The trends for LS vs LF as well as LF vs DF are also positive and almost linear (Fig. 12.7C, D). However, the values are more scattered for the trends and are indicated by thinner and thin lines in Fig. 12.6C, D. The values for fecundity vs GT as % of LS are even more scattered (Fig. 12.7A). Nevertheless, the values for free-living worms fall consistently at lower levels than those for parasitic worms. Clearly, (1) fecundity is more a function of body size than LS or GT % of LS. (2) Parasitic worms are more fecund than free-living worms. (3) Among vertebrate parasites, the oviparous nematodes involving no intermediate host are more fecund than the viviparous nematodes, whose transmission involves hematophagous host.

At this point, it must also be noted that inequality in lifetime fecundity has been described for many parasitic helminths. For example, just 10% of the pike cestode *Trianophorus crassus* population produce ~ 85% of the eggs

(Shostak and Dick, 1987). Gini coefficient and Lorenz curve are used to measure extreme Variations in Reproductive Success (VRS, Hanelt, 2009) or Inequality in Lifetime Fecundity (ILF). A value of 0 (the trend for equality) indicates a perfect reproductive equality among females in a population but 1 shows that reproduction is limited to a few dominant individuals within a population (Hanelt, 2009). The variable Gini coefficients reveal that only a few dominant females contribute to reproduction at population level in some nematodes; for example, the Gini coefficients are 0.43 for a mermithid (Poulin and Lotham, 2002), 0.56 for an the ansakid *Rhabidascaris acus* (Szalai and Dick, 1989) and 0.40–0.58 for *Hymenolepis diminuta* (Dobson, 1986). These values clearly indicate that ~ 50% parasitic nematode females contribute more to reproductive output than others. With gonochorism, only that female along with a male in the same host individual may contribute to reproduction.

Embryogenesis and Life Cycle

The description on embryogenesis is based on Hyman (1951b) and Hope (1974). In some nematode species, the zygote commences development *in utero* (e.g. *Rhabditis marina, Monhystera disjuncta*) and in others, its *in utero* development may lead to ovoviviparity and viviparity. But in most others, it commences only after oviposition. In the oval egg, anterior pole is marked by the micropyle, through which the sperm enters. The first cleavage is transverse to the egg axis and produces more or less equal anterior blastomere called S1 and a posterior blastomere P1, from which the germ cells descend (Fig. 12.8A, B). Hence, the cleavage almost establishes the germ and somatic cell lineages even at the very first division. At the second cleavage, S1 divides into A and B blastomeres; P1 divides into S2 and P2 blastomeres. In this rhomboid, A is anterior, B dorsal, S2 ventral and P2 posteior. The A and B give rise to the ectoderm of the entire body, except the posterior end. The S2 cell called the ectodermal mesodermal somatodeal (EMSt) divides into E and MSt. The descendants of E give rise to endoderm and the MSt the primary mesoderm and somatodaeum. In the P lineage, (a) P2, (b) P3 and (c) P4 gives rise to (i) secondary ectoderm and tertiary mesoderm, (ii) secondary mesoderm and (iii) Primordial Germ Cells (PGCs), respectively (Fig. 12.8B). Experiments have confirmed the highly determinate mosaic nature of development.

Prior to the description of life cycle, an explanation on dormancy and encystation have to be provided. To survive adverse conditions, some nematodes have evolved an effective dormant stage, which extends their life span. During dormancy, the nematodes may not feed but suppress metabolism to a low level. Usually, dormancy is induced during the J3 stage. In *Caenorhabditis elegans*, induction of dormancy is regulated by starvation,

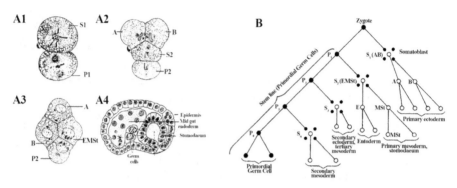

FIGURE 12.8

A. Cleavage and gastrulation stages in development of nematode (redrawn from Nigon, 1965).
B. Cell lineage in a nematode (redrawn from Tadano, 1968).

high temperature and/or high population density. The decision to enter into dormancy is controlled by a number of hormonal signaling (Sommer, 2015). Encystation is a strategy to extend the duration of a specific infective stage. For example, J4 of *Trichinella spiralis* coils up and remains quiescent, and the host forms a cyst around it that is eventually calcified. Not much is known about the energy source for the sustanence of the dormant or encysted state.

As mentioned earlier, life cycle of nematodes involves six stages: the egg, four juvenile and adult stages. In free-living nematodes, it is simple and direct. It is, however, diverse and complicated in parasitic nematodes. In mermithids, juveniles are parasitic, while their adults are free-living. In non-mermithid parasitic species, adults are parasitic and their juveniles may be free-living. The carcass-feeding mermithids may kill the host (e.g. *Steinernema*, *Heterorhabditis*, *Ressimermis* mosquitoes) (Fig. 12.9B), but feeding on hemocoel and fat bodies may render the host to survive but with considerably reduced fecundity (see also grasshopper *Melanoplus dawsoni*, Fig. 12.5A). The duration of parasitic phase may last from 7–12 days in *Hexamermis arvalis* (Poinar and Gyrisco, 1962) to 395 days in *Agamermis decaudata* (Table 12.12). A rare publication reports that *Pheromermis pachysoma* completes parasitic juvenile phase in caddisfly, which is parasitized by a wasp, a second transport host. From the wasp, *P. pachysoma* adults emerge (Fig. 12.9C). All tetradonematids, for example, *Tetradonema plicans* complete its entire parasitic life (both juvenile and adult stages except for the egg stage) in *Sciara* fly (Fig. 12.9A).

In non-mermithid nematode parasites, the cycle may be direct but mostly involves one or two host(s). Further, the cycles are diverse and complicated to extend the life span by (i) inclusion of Extra-Intestinal Migration (EIM), (ii) insertion of dormant and/or encysted life stage(s) or (iii) interspersion of parthenogenesis. In man, the intestinal round worm *Ascaris lumbricoides* and hookworm *Ancylostoma duodenale* pass through a simple direct life

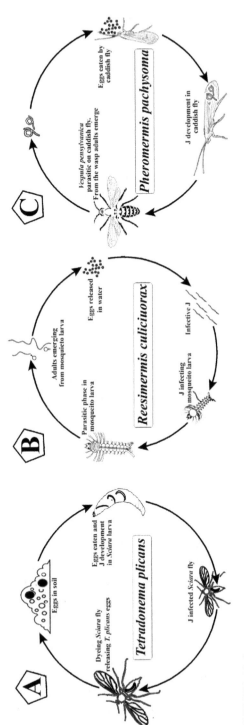

FIGURE 12.9

Life cycle of some mermithids: A. the tetradoenematid *Tetradonema plicans*: note the entire life cycle is completed after killing the larva and adults of *Sciara* fly, B. the mermithid *Reesimermis culiciuorax*: note that the adults are free-living and C. *Pheromermis pachysoma*, in which juvenile phase of life cycle is completed in caddisfly, which is parasitized by a wasp, from which free-living adults emerge (drawn from the description of Hyman, 1951b, Petersen, 1975 and Poinar et al., 1976).

cycle with no intermediate host but extends their life span by undertaking an EIM (Fig. 12.11A). In the hookworms, the EIM damages a number of organs and systems more severely than their presence in the intestine. The rhabditoid *Neoaplectana glaseri* includes repeated parasitic cycles as long as food is available in the carcass of the infected Japanese beetle *Popillia japonica*. Following a free-living life for 18 months, its J3 infects a fresh beetle grub to continue its adult parasitic life (Fig. 12.10B). Depending on environmental conditions, the tylenchid *Bursaphelenchus xylophilus* may choose one of the following: (i) propagatory, (ii) dormant or (iii) dispersal cycle. In the propagatory cycle, the worm directly completes its cycle within 25 days but in the second, its J3 undergoes dormancy lasting for several months in pupal chamber of the beetle *Monochumus*. Subsequently, it molts into J4 and the adult stage. In the third dispersal cycle, J4, after the dormancy and subsequent molt from a pupal chamber, enters an adult beetle through the spiracle into tracheal system and awaits its dispersal among the pine foliages. At an appropriate opportunity, it molts into the adult stage (Fig. 12.10C).

Intermediate hosts: The cycle may involve one host in the spectacular human helminth parasites *Dracunculus medinensis* and *Wuchereria bancrofti*. In the former, the J2 is transmitted by *Cyclops* and the infection is through drinking the *Cyclops*-contaminated water (Fig. 12.11A). In the latter, it is uniquely by blood sucking mosquitoes (Fig. 12.11B), biting midges, blackflies or mites. In still others, dioctophymoid *Dioctophyme renale* and spiruroid *Spiroxys contortus*, the cycle involves two intermediate hosts (Fig. 12.11C, D). The inclusion of alternate intermediate hosts reduces the risk involved in transmission events from definitive to First Intermediate Host (FIH) and FIH to Second Intermediate Host (SIH). For example, *S. contortus* may choose one among the five *Cyclops* species as FIH and one among the six fish, amphibians and dragonfly nymph species for transmission from FIH to SIH. To complete one generation of life cycle *D. renale* requires a minimum of two years (Hyman, 1951b). Hence, the involvement of one or two intermediate host(s) extends the life span of these parasitic nematodes.

Parthenognesis: In yet other parasitic nematodes, the cycle may involve homogonic parthenogenesis alone (e.g. *Strongyloides venezuelensis*, Hino et al., 2014) or heterogonic parthenogenesis involving repeated parthenogenesis interspersed by sexual reproduction (e.g. *S. ratti*, Hyman, 1951b). For example, the 2 mm long tylenchid *Deladenus siricidicola* undergoes repeated parthenogenic cycles feeding on a symbiotic wood-rotting fungus *Amylostereum areolatum* inhabiting the pine tree (Fig. 12.12B). According to Hajek and Eilenberg (2018), the presence of high CO_2 and low pH induce an inhabitant females of *D. siricidicola* to produce infective females. The 1 mm long parasitic female is infected by the *Sirex noctilio* larva. The parasitic female grows 1,000-times and releases juveniles, which invade *S. noctilio* reproductive system. Eventually, the infected *S. noctilio* adult lays its eggs along with the nematodes into a fresh pine tree.

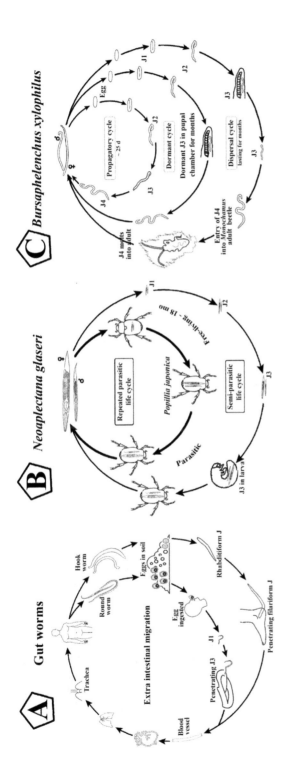

FIGURE 12.10

A. Life cycle of *Ascaris lumbricoides* and *Ancylostoma duodenale*. B. *Neoplectana glaseri*: repeated propagatory parasitic life cycle in *Popillia japonica* and semi-parasitic cycle involving free-living juvenile stage for 18 months (drawn from the description of Hyman, 1951b). C. *Bursaphelenchus xylophilus*: the propagatory, dormant and dispersal life cycle (free hand drawings, modified from Zhao et al., 2013).

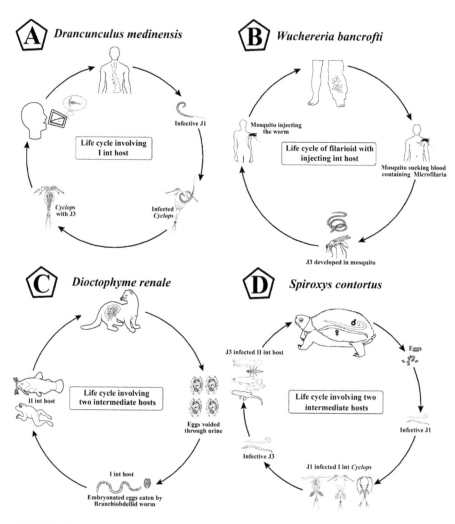

FIGURE 12.11

Life cycle of parasitic nematodes involving one intermediate host in A. *Dracunculus medinensis* and B. *Wuchereria bancrofti* (drawn from description of Hyman, 1951b), C. *Dioctophyme renale* (free hand drawing, modified from Hyman, 1951b, Global Health: Parasitic Diseases) and D. *Spiroxys contortus* involving two intermediate (int) hosts (drawn from description of Hedrick, 1935).

The rhabdiasoid *S. ratti* includes both the homogonic and heterogonic strains. Each strain produces predominantly its own type. The yield of heterogonic from homogonic is related to the season, being greater in spring and summer than in autumn and winter (Fig. 12.12A). However, the cycle is limited to homogonics alone in *S. venezuelensis*. A karyotype study has revealed the presence of 2 pairs (2n = 4) of chromosomes in *S. venezuelensis*

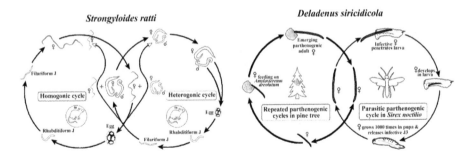

FIGURE 12.12

Homogonic and heterogonic life cycle of *Strongyloides ratti* parasitic on rat. Note that a small proportion of homogonic cycle produces male and female progenies and thereby minimizes inbreeding depression (drawn from description of Hyman, 1951b). *Deladenus siricidicola*: Free-living life cycle on pine tree and parasitic life cycle on *Sirex noctilio* (free hand drawing, modified from description of Hajek and Eilenberg, 2018).

but 3 pairs including one sex chromosome in *S. ratti*. In *S. venezuelensis*, the length of two chromosomes significantly differ from others and the long one is suggested to result from a fusion of the sex chromosome with another autosome (Hino et al., 2014). Hence, the presence of heterogonic and homogonic parthenogeneses has a genetic base in parasitic nematodes but its expression may be modulated by environmental factors.

Sex Determination

A vast majority of nematodes are gonochores. However, the sporadic incidence of protandric hermaphrodism is not uncommon across the taxonomic groups. Different types of parthenogeneses namely (i) amictic/ameiotic, (ii) mictic/meiotic/pseudogamic (e.g. *Rhabditis monhystera*) and (iii) haplo-diploid/generative (e.g. parasitic oxyuroids, Morand et al., 2005) parthenogeneses are reported. It is also known that parasitic load or intensity significantly alters sex ratio in some parasitic nematodes. For example, sex ratio of *Mermis subrigrescens* remains equal so long the intensity is < 3 worm/grasshopper. But it is skewed in favor of males at 3–14 worm/grasshopper. When the intensity increases to more than 14 worm/grasshopper, only males are developed (Hyman, 1951b). This is also true of plant parasite *Meloidogyne incognita* (Hope, 1974). In most nematodes, sex determination is strictly genetic but it may be primarily genetic with environmental (parasitic intensity) modulation at the downstream levels of genetic cascade of sex differentiation (Pandian, 2011).

The first discovery of sex chromosome was in *Mesorhabditis belari*, in which one sex chromosome is larger than autosomes but the other is as short as autosomes. In *Contraecum incurvum*, the identified Y-chromosome is paired with a long X-chromosome. However, there are also 7 univalent X-chromosomes (see Hope, 1974). Almost all the known mechanism of sex determination have shown male heterogamety in nematodes, which can be XX-XY type, as in *Anguina tritici, Burgia malayi, Baylisascaris trausfuga, C. incurvum, Meloidogyne hapla, Onchocera volvulus, Trichurus muris* (Hope, 1974, Morand et al., 2005) or XX-XO type, as in *Haemunchus contorturs, Strongyloides ratti, Trichinella spiralis* (Morand et al., 2005). In the presence or absence of Y-chromosome, the male sex is determined but in those with XX chromosomes, the female sex is determined. In the protandric hermaphrodite *Caenorhabditis elegans*, the ratio for X-chromosomes to autosomes controls a complex signaling pathway of negative active factors. During the male phase, a master sex determination switches the gene *xol-1* to masculinization. In the female phase, the suppression of *xol-1* activates the zinc transcription factor *TRA-1* and thereby suppresses the masculanization and lets the feminization process to go on (Sommer, 2015).

13

Nematomorpha

Introduction

Being eutelic (Blackwelder and Garoian, 1986) and pseudocoelomate, the nematomorphs are a sister phylum to the Nematoda. Though sharing many features such as slender body shape, cuticular organization and so forth, the following key features delineate them as a distinct phylum: (1) They are all parasitic as juvenile on arthropod hosts and free-living as adults. (2) In them, the gut is absent or non-functional, when present. (3) Hence, the juveniles feed osmotrophically on the host's hemocoel. (4) The surface of their epicuticle is criss-crossed with furrows and elevations of irregular area called 'areoles' (Hyman, 1951b, Hanlet et al., 2012) and thereby increases the surface area for osmotrophic feeding. Their color ranges from yellowish to mostly black and size measures from 15–25 cm in length to 2 m (e.g. *Gordius fulgur*, Fig. 13.1D) and 0.25–3.0 mm width (Sasi and Giannetto, 2016).

Taxonomy and Distribution

The phylum Nematomorpha comprises two classes: The freshwater Gordian worms Gordiaceae consisting of 12 genera and the marine hairworms Nectonematoidea with a single genus *Nectonema* (Fig. 13.1E) including five species. A conservative estimate for the global species number is 2,000 (Poinar, 2008). However, the number of described species ranges from 300 (Schmidt-Rhaesa, 1997) to 326 (Poinar, 2008) and to 360 (Schmidt-Rhaesa, 2013), indicating that the average increase in the number of species description is increased at the rate of 3.8 species/y. *Nectonema* are parasites of decapods crustaceans (Sasi and Giannetto, 2016). For freshwater nematomorphs, a wide range of arthropods, especially orthopterans and coelopterans, and others serve as hosts.

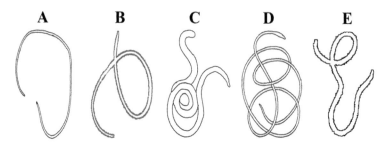

FIGURE 13.1

A. *Gordius* sp (Univ Florida), B. *Chordedes japonensis* (from Present5), C. *Paragordius tricuspidatus* (from Present5), D. *Gordius fulgur* (from biologydiscussion), E. *Nectonema agile* (free hand drawing).

Reproduction and Life Cycle

A vast majority of nematomorphs are gonochores and reproduce sexually. As in nematodes, the males can be distinguished from their smaller size (e.g. *Gordius robustus*, Hanlet and Janovy, 1999). Their gonads consist of a pair of long cylindrical bodies, whose respective ducts are extended, and open separately into the cloaca. The sperm are deposited as spermatophore on the female terminus or as spermatozoa directly into the cloaca. Their life cycle includes four stages: (i) the egg, (ii) the pre-parasitic larva, that hatches from the egg and is quite different from the adult, (iii) the parasitic juvenile and (iv) the free-living aquatic adult male and female. Their indirect life cycle may involve no Intermediate Host (IH) or one IH. For example, Hanlet and Janovy (1999) experimentally demonstrated that *G. robustus*, a parasite on *Gryllus firmus*, does not require an IH. Schmidt-Rhaesa et al. (2009) hinted at that many nematomorphs complete their entire life cycle within aquatic system involving caddisflies or diving beetles as DH (Fig. 13.2A). For a list of other arthropods serving as DH, Cappucci (1976) may be consulted. On the other hand, the gordiids like *Chordodes japonensis* (Fig. 13.1B) do require an IH. Inoue (1962) showed that 21 of 26 (81%) mantids became infected, when fed the mayfly *Cloeondipterum* containing cysts of *Ch. japonensis* (Fig. 13.2C). It is difficult to comprehend how the emerged adults from the mantids and crickets can reach their aquatic 'home'. From their laboratory and field experiments, Thomas et al. (2002) indicated that the nematomorph *Paragordius tricuspidatus* (Figs. 13.1C, 13.2B) manipulates the infected cricket *Neonobius sylvestris* to jump into water so that the adults could emerge in aquatic habitats. The need for linking the life cycle of aquatic adults and terrestrial orthopterans and vice versa is obvious. It is difficult to comprehend how the terrestrial cricket like *N. sylvestris* is infected by the aquatic *P. tricuspidatus*, albeit the infected cricket is driven to dive into

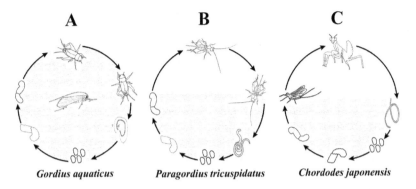

A   **B**   **C**

Gordius aquaticus   *Paragordius tricuspidatus*   *Chordodes japonensis*

FIGURE 13.2

Life cycle of selected nematomorphs. The cycle includes the egg, pre-parasitic and penetrative larval stages and endoparasitic juvenile phase. From the host, adults enter the aquatic habitat. A. *Gordius aquaticus*, in which the life cycle is completed entirely in the aquatic habitat, involves no intermediate host. In the place of the diving beetle, a caddisfly also can serve as a definitive host. B. *Paragordius tricuspidatus*, in which the larval and adult stages are aquatic but the juvenile parasitic phase is developed in terrestrial cricket host. C. *Chordodes japonensis*, in which the larval and adult stages are aquatic but an infected mayfly serves as an intermediate host to transfer the parasite from aquatic to terrestrial definitive host.

water. The terrestrial mantid, that ingests the infected mayfly *Cl. dipterum* emerging from an aquatic habitat, may readily be infected with *Ch. japonensis* but it is difficult to comprehend how the adult parasite returns to the water, when emerging from the mantids. It is likely that mantids and crickets that inhabit aquatic plants emerging from water levels are alone engaged as hosts. A number of publications report on the discovery of new host like the diplopod *Cambala annulata* serving as host to *Gordionus* (Schmidt-Rhaesa et al., 2009) but it has not been identified whether they are IH or paratenic host (PH). Yadav et al. (2017) reported the presence of encysted *Gordius* sp in pork; apparently, the pig serves as PH. The involvement of a large number of IH and PH indicates that nematomorphs are extremely generalists and have the capacity to encyst in a variety of hosts and some of which may come in contact with an aquatic system (Hanlet and Janovy, 1999).

Lifetime Fecundity

Hanlet (2009) is perhaps the only contribution that provides an idea about some life history traits and Lifetime Fecundity (LF) in the gordid worm *Paragordius varius* reared on cricket *Gryllus firmus*. From his experimental study, the following may be inferred: (1) During the parasitic juvenile phase lasting for 25–30 days, the worm grows to a maximum length of 34 cm and turns its body color from light yellow to black. (2) Females survive longer than the males (Fig. 13.3A). Frequency of the infected worm increases up to 6 at the parasitic intensity of 6 worm/host and subsequently decreases to 1 beyond 16–30 worm/host (Fig. 13.3B). (3) Cumulative length of the worms in a host increases directly and linearly up to the intensity of 30 worm/host for females but levels off beyond 12 worm/host for male (Fig. 13.3C). (4) However, the length of the individual worm decreases with increasing intensity, albeit at a higher level for females than that for males (Fig. 13.3D). (5) Lifetime fecundity increases from 100,000 eggs for a worm measuring 14–16 cm in length to 5.8 million eggs in the lengthiest worm of 34 cm (Fig. 13.3E). The correlation between worm length and LF is significant and explains much of variations in LF. (6) Analysis of Gini coefficient (0.2) and Lorenz curve for *P. varius* (Fig. 13.3F) and *P. difficilis* (from a field study, not shown in figure) suggest that within populations, the offspring of next generation contributed almost equally by females in these Gordiids.

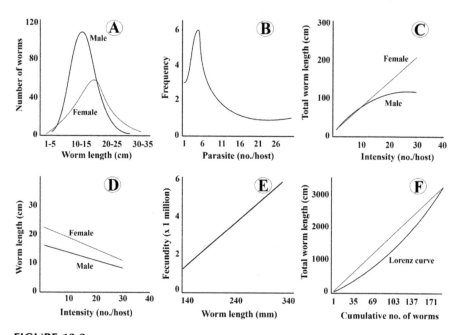

FIGURE 13.3

Paragordius varius: A. Frequency distribution of males and females as a function of body size. B. Frequency distribution as a function of intensity. C. Cumulative length of worms as a function of parasitic intensity. D. Individual worm length as a function of parasitic intensity. E. Fecundity as a function of body size. F. Cumulative worm lengths plotted against cumulative number of worms. Lorenz curve drawn on the basis of Gini coefficients (modified and redrawn from Hanelt, 2009).

Parthenogenesis

Whereas a majority of nematomorphs reproduce sexually, some are parthenogenics. Of 18 *Paragordius* species examined (Table 13.1), males are known only in 7 (39%) species. Of 9 *Paragordius* species reported from Africa, the incidence of males has been documented for only one species *P. cinctus*. Eliminating the involvement of *Wolbachia* and similar bacteria that kill males alone (Pandian, 2016). Hanlet et al. (2012) showed that 11 of 18 (61%) *Paragordius* species may undergo mitotic parthenogesis.

TABLE 13.1

Distribution and presence of males in *Paragordius* spp (modified from Hanelt et al., 2012)

Geographic Region	Species
Species, in which males are not found	
Africa	*P. areolatus* *P. dartevellei* *P. laurae* *P. marlieri* *P. obamai* n. sp *P. samaliensis* *P. tanganikensis*
Central America	*P. diversolobatus* *P. flavescens*
South America	*P mulungensis*
Oceania	*P. emeryi*
Species, in which males are recorded	
Africa	*P. cinctus*
Europe	*P. stylosus* *P. tricuspidatus*
America	*P. andreasii* *P. varius*
South America	*P. esavianus* *P. minusculus*

14

Acanthocephala

Introduction

The 1,100 speciose (Nickol, 2006) Acanthocephala comprise eutelic (Whitfield and Evans, 1983), pseudocoelomatic, obligate endoparasites that engage arthropods and vertebrates in a conserved two-host life cycle (Near, 2002). In them, the body is divisible into a short slender prosoma, a neck with no hooks and a much longer stouter trunk. The diagnostic feature of the phylum is the organ of attachment consisting of an anterior invaginable proboscis armed with rows of recurved hooks. The trunk may be cylindrical, flattened, curved or coiled with a smooth, wrinkled or segmented surface (Fig. 14.1). The body is clothed with a thin (1 μm) cuticle, epidermis, dermis and muscle stratum. Beneath the homogenous cuticle lies a thick layer of fibrous syncytial epidermis. The muscle layer consists of permeable outer circular and inner longitudinal fibres. The mouth, anus and digestive tube are absent. So is the circulatory system. The nervous system consists of a ganglion near the proboscis and two lateral chords. The hollow interior is occupied by the reproductive system. The ligament sacs, a peculiar feature

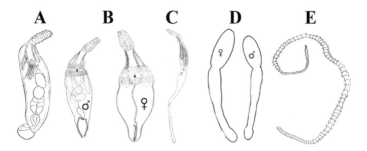

FIGURE 14.1

A. *Acanthocephalus renae* (after Von Cleave, 1915). B. *Andracantha tandemtesticulata* (from Monteiro et al., 2006). C. *Arhythmorhynchus* (after Luhe, 1912). D. *Profilicollis altmani* (from wikidictionary). E. *Mediorhynchus taeniatus* (after Meyer, 1933).

of Acanthocephala, are hollow tubes running through the length of the body enclosing the reproductive organs (Hyman, 1951b). A majority of acanthocephalans measure < 25 mm in length but there are a few measuring 65 cm (e.g. *Macroacanthorhynchus hirudinaceus*). They are colorless but with absorption of food, especially fatty substances from the host, they may be red, orange, yellow or brown in color.

Taxonomy and Distribution

The number of described acanthocephalan species increased from 1,000 (Kennedy, 1993) to 1,100 (Nickol, 2006). The rate of increase in new species erection was 8.3 species/y between 1993 and 2006. At this rate, the total number of acanthocephalan species may not go beyond 1,250. These species are classified into four classes, which include nine orders (Table 14.1). The class Polyacanthocephala consists of only four species. With the exception of nine species found in North American turtles, eoacanthocephalans are

TABLE 14.1

Classification and characteristic features of Acanthocephala (condensed from Nickol, 2006)

Orders	Characteristic Features	Definitive Host	Intermediate Host
Polyacanthocephala			
Polyacanthorhynchida	Dorsal, ventral main lacunar canals	Crocodiles	Not known
Archiacanthocephala			
Apororhynchida	Globular, non-retractable proboscis	Aves, mammals	Terrestrial insects
Gigantorhynchida	Conical retractable proboscis	Aves	Not known
Moniliformida	Cylindrical retractable proboscis	Mammals	Cockroaches
Oligacanthocephala	Spherical retractable proboscis	Mammals, aves	Orthoptera, Coeloptera
Palaeacanthocephala			
Echinorhynchida	Infects lower vertebrates	Fish, amphibians	Amphipods, isopods
Polymorphida	Infects higher vertebrates	Aves, mammals	Amphipods, isopods, decapods
Eoacanthocephala			
Gyracanthocephala	With trunk spines	Fish	Copepods
Neoechinorhynchida	Without trunk spines	Fish	Ostrocods

parasites of fishes. Most of them are found in the cypriniform families Cyprinidae and Catostomidae than in any other piscine family (Nickol, 2006). They are mostly cosmopolitan but a few are endemic. Within the genus *Macroacanthorhynchus, Ma. hirudinaceus* are ubiquitous throughout the world. *Ma. ingens* is endemic to America and *Ma. catulinus* to Asia (Saari et al., 2019). In the absence of a gut, they are relegated to the full length of the small intestine but are mostly found on the posterior part of the anterior intestine of the host (e.g. *Moniliformis moniliformis,* Crompton et al., 1988). However, site selection within the intestine modulates the incidence, intensity, male-male competition and so forth (Sinisalo et al., 2004).

Acanthocephala are gonochores with pronounced sexual dimorphism; the females are larger than the males from the juvenile cystacanth phase to the adult stage. The sexual size dimorphism, as estimated by the number of times the female is longer than the conspecific males, ranges from 1.08 times for cystacanth and 1.11 times for adult of *Neoechinorhynchus rutili* and from 2.08 to 2.77 times for cystacanth and adult *Acanthocephalus parkseidi,* respectively. Benesh and Valtonen (2007) have assembled these values for 14 species. These values average to 1.4 times for the cystacanth and 1.9 times for the adults. Hence, the level of sexual size dimorphism is magnified during development onwards.

The foregone description indicates that in them, sex is determined early at the time of fertilization. Not surprisingly, the sex ratio, for example in *Echinorhynchus truttae* cystacanth, remains at 0.53 ♀ : 0.47 ♂, irrespective of changes in intensity of infection from 60 cystacanth/*Gammarus pungen spadanus* to 1,085 cystacanth/host (Parenti et al., 1965), clearly indicating the existence of a rather rigid system of sex determination. In fact, male heterogamety is reported for XX-XY system in *Ma. hirudinaceus* and XX-XO system for *Mo. moniliformis* (Robinson, 1965). Still, the ratio may slightly be skewed in favor of females. This is due to the fact that the female, even during the cystacanth stage, survives longer than the male (e.g. *Acanthocephalus lucii, Echinorhynchus borealis,* Benesh and Valtonen, 2007).

Reproductive system: In males, the system consists of a pair of testes, a small penis, cement glands and copulatory bursa (Fig. 14.2). In females, the original one or double ovary breaks up into the fragmented ovarian balls that float freely in liquid of the body cavity (Fig. 14.2). The eggs in these balls are fertilized following copulation. They are collected by a funnel-shaped uterine bell, which screens and lets only the mature embryonated eggs to be voided into the uterus, from where they are eventually released into the exterior through an oviduct. Unlike the nematodes, in which the sperm is a limiting factor of fertility, the acanthocephalans are not limited by the sperm. In them, no hermaphroditism is thus far reported and they do not reproduce parthenogenically or clonally.

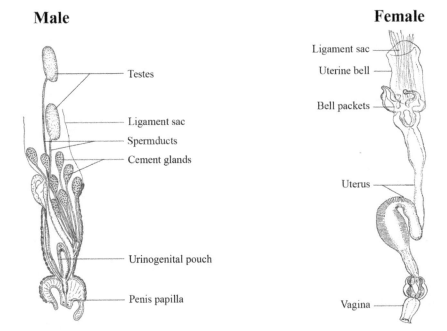

FIGURE 14.2

Reproductive system of male *Hamanniella* (after Kilian, 1932) and female palaeacanthocephalan *Bolbosoma* (after Yamaguti, 1939).

Development and Life Cycle

Development: According to Nicholas and Hynes (1963), the cleavage pattern in the acanthocephalan *Polymorphus minutus* is a distorted spiral type. The egg cleaves into four blastomeres located linearly (obligue to each other in *Macroacanthorhynchus hirudinaceus*, Hyman, 1951b), instead of the usual paralleled position (Fig. 14.3). All subsequent divisions are not synchronized and are unequal. Following several generations of division, the macromeres and micromeres become similar in size. However, the embryo is now more a syncytium in its organization. Some nuclei of micromere origin are located at the anterior and posterior positions; they do not condense but generate cortical nuclei. Eventually, the central nuclear mass differentiates into all adult organs except the body wall. For an easier understanding, organogenesis is illustrated for *Ma. hirudinaceus*. Notably, primordium for the gonad and ligament is identifiable at this stage.

Life cycle: In Acanthocephala, the cycle is indirect and involves arthropods or vertebrates as Intermediate Hosts (IHs) and vertebrates as Definitive Hosts (DH). In aquatic habitats, crustaceans like amphipods, isopods, copepods,

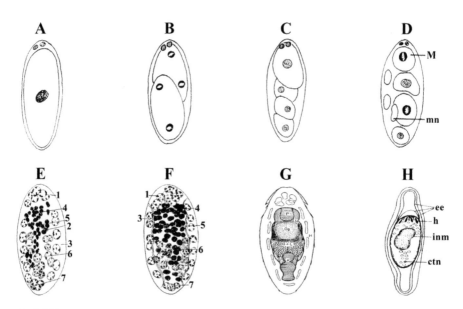

FIGURE 14.3

Early cleavages (A–D) and acanthor (H) in *Polymorphus minutus* (redrawn from Nicholas and Hynes, 1963). Organogenesis (E–G) in *Macracanthorhynchus hirudinaceus* (redrawn from Mayer, 1936). M = macromere, nm = macromere nucleus, ee = egg envelope, h = hook, inm = internuclear mass, ctn = cortical nucleus, 1 = proboscis primordium, 2 = condensed nuclei, 3 = definitive epidermis, 4 = brain primordium, 5 = musculature primordium, 6 = primordium of the gonads and ligament, 7 = primordium of the urinogenital system except gonads.

ostrocods and others serve as IH (Table 14.2). Schmidt (1985) provides a list of them. The embryonated eggs released from DH contains the first larva, the acanthor (Fig. 14.4 A). On ingestion of the egg by an IH, the acanthor develops into an acanthella. During these stages, primordia for all organs are laid and developed. Late in acanthella development, the proboscis invaginates and all structures of adult worm are formed. This juvenile phase called cystacanth is infective to DH (Nickol, 2006). Briefly, acanthocephalans pass through two larval stages and a juvenile stage in IH. Hence, the worm may reduce survival, growth and reproduction of IH. However, no information is yet available on this aspect. It is also notable that the only trophic mode of transmission occurs from DH to IH and IH to DH. When a cystacanth is eaten by a non-appropriate host, it remains as cystacanth in the paratenic host. Small fishes are important paratenics for many species, especially *Corynosoma semerme*, to whom marine birds and mammals are the DHs (Nickol, 2006). For freshwater *Pomporhynchus leavis*, small cyprinids serve as paratenics (Mehlhorn, 2016). For terrestrial thorny-headed worms also, the paratenics have been identified. For example, the adult beetle and lizard may serve as paratenics for *Macroacanthorhynchus* spp (Saari et al., 2019).

TABLE 14.2

Definitive (DH), intermediate (IH) and paratenic (PH) hosts of some Acanthocephala (condensed from Mehlhorn, 2016)

Parasite	DH	IH	PH
Archiacanthocephala			
Moniliformis moniliformis	Rodents	Cockroaches	-
Macaracanthorhynchus hirudinaceus	Pigs	Beetle larvae	-
Ma. ingens	Raccoons	Beetles	Amphibia, Reptilia
Prosthenorhcis elegans	Monkeys	Cockroaches, beetles	-
Palaeacanthocephala			
Acanthocephalus anguillae	Cyprinids	*Asellus aquaticus*	Small cyprinids
A. renae	Amphibia	*A. aquaticus*	-
Echinorhynchus truttae	Salmonids	*Gammarus*	-
Corynosoma semerme	Seals, birds	*Pontoporeia affinis*	Various marine fish
Pomphorhynchus laevis	Cyprinids	Gammarids	Cyprinids
Filicollis anatis	Ducks, aquatic birds	*A. aquaticus*	-
Polymorphus minutus	Ducks, aquatic birds	Gammarids	-
Eoacanthocephala			
Neoechinorhynchus cylindratus	Predatory fish	*Cypria globula*	Bluegills
N. emydis	Turtles	*C. globula*	Aquatic snails
N. rutili	Salmonids	Ostrocoda	-
Paratenuisentis ambiguus	Eels	*Gammarus tigrinus*	-

As in cestodes, the life cycle for many acanthocephalans remains to be described. For example, the cycle of *Acanthocephaloides propinquus* involves *Gobius buchichii* to serve as DH. However, the IH is not identified (Sasal et al., 2000, 2001). Mehlhorn (2016) lists DH, IH and PH for only 17 important acanthocephalan species. The list suggests specificity for IH in Palaeacanthocephala and Eoacanthocephala but not for DH (Table 14.2). In Fig. 14.4, an attempt has been made to describe three types of life cycles. Accordingly, the cycle of *Corynosoma cetaceum* involves the franciscana dolphin *Pontoporia blainvillei* and the striped weakfish *Cynoscum gnatucupa* or the rough scad *Trachurus lathami* serve as IH (Fig. 14.5A). To *Acanthocephalus anguillae*, the cyprinid *Squalius cephalus* and the isopod *Asellus aquaticus* serve as DH and IH, respectively (Fig. 14.5B). In terrestrial habitat, the beetle larva and dogs serve as IH and DH for *Macroacanthorhynchus hirudinaceus*. Adult beetles and lizards may serve as PH. The limited relevant information hints that the IH size seems to limit LF of the paratenic worm. For example, LF

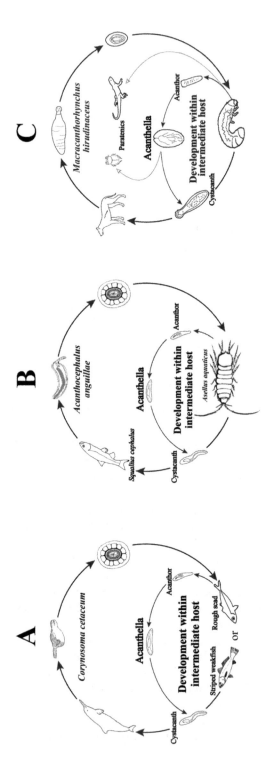

FIGURE 14.4

Life cycles in Acanthocephala: A. *Corynosoma cetaceum* involving vertebrate intermediate host (based on Aznar et al., 2018). B. *Acanthocephalus anguillae* involving invertebrate intermediate host (based on Mehlhorn, 2016). C. *Macracanthorhynchus hirudinaceus* involving insect invertebrate host. In paratenics, development of cystacanth does not proceed (based on Saari et al., 2019).

runs from 11,000 eggs to ~ 90,000 for *C. cetaceum* (Fig. 14.5A), which involves the large striped weakfish (1 m, 9 kg) or rough scad (225 mm) as IH. But it ranges from up to 12,000 eggs for *Acanthocephaloides propinquus* (Fig. 14.5E) engaging *Asellus aquaticus* (maximum size 10 mm length). Of course, the DH dolphin is far larger than the DH *G. buchichii*. Hence, there may be a size relation between DH and IH as well as LF.

Fecundity

Prior to description of fecundity, a prelude on a few characteristic features of acanthocephalans becomes necessary: (1) *Immunity*: (A) The intermediate arthropod hosts may not be as immune as the definitive vertebrate hosts are. (B) Within vertebrate hosts also, the duration and level of immunity development may vary from one taxonomic group (e.g. birds) to another (e.g. immunization via milk in mammals). (2) *Quantity and quality of food* to the host has a profound effect on survival, growth and reproduction of the worms. (A) For example, natural starvation eliminates *Profilicollis botulus* from the infected elder duck *Somateria mollissima* (Thompson, 1985). Experimental starvation eliminates *Mo. moniliformis* from rats (Burlingame and Chandler, 1941) and *Polymorphus minutus* from domestic ducks (Nicholas and Hynes, 1958). Due to its availability and amenability to culture, more information is available on rats infected with *Mo. moniliformis*. In *Mo. moniliformis*, a carbohydrate-free diet may facilitate its establishment in rats but not its survival, growth and reproduction in surviving rats. In rats fed on glucose diets, parasitic growth is directly related to the glucose concentration in the host diet (Crompton et al., 1982). Feeding the rats with 3 or 6% fructose, the parasites were larger, contained more ovarian balls and produced more number of eggs than rats fed on 1% fructose diet (Crompton et al., 1988). In fact, rats fed on a low fructose (< 3%) diet, the number and size of ovarian balls decrease and only a few or no eggs are produced (Keymer et al., 1983). Dobson (1986) demonstrated that the dietary fructose dose and parasitic intensity significantly modulate Gini coefficients. For example, the coefficient decreases from 0.7 at 1% fructose diet to 0.3 at 12% fructose diet in rats infected with 80 worm/host; however, the decrease is from 0.5 to 0.2 in rats infected with 40 worm/host. (3) *Social influence*: Acanthocephalan males are polygamous; a male is capable of inseminating several females. Young males can inseminate old ones. But sexual contact declines with age (Parshad and Crompton, 1981). The site selection in the intestine may increase mating frequency (Crompton, 1976). (4) *Sperm availability*: (A) Insemination is required at least once every five weeks to provide sperm and to ensure fertilization (Crompton, 1985), suggesting the existence of batch fecundity in *Mo. moniliformis*. (B) Degeneration of mature oocytes occurs more frequently in *Centrorhynchus corvi* females that have not been inseminated (Parshad and Guraya, 1978), clearly indicating the existence of the potential and realized fecundity in *C. corvi*. (5) *Seasonal periodicity*: In poikilothermic definitive hosts like fish differ from homeiothermic birds and mammals. In temperate zones, temperature plays an important role in establishing seasonal reproductive periodicity. A common cycle consists of (i) recruitment of fish in the gradually warming spring, (ii) both fish and IH becoming active during summer, facilitating the infection of the parasite from DH to IH, (iii) slow development of the parasite during autumn and completion of larval development to the infective cystacanth stage at the end of winter (Nickol, 2006). For example, seasonal differences in densities of *Echinorhynchus truttae* cystacanth results on seasonality, although the IH *Gammarus pulex* is available to *Salmo trutta* almost throughout the year (Awachie, 1966). In others like *Fessisentis fuedi* in *Esox americanus*, *Pomphorhynchus bulbocolli* in *Catastomus commersoni* and *E. salmonis* in *Oncorhynchus tshawytscha*, the periodicity is related to migration

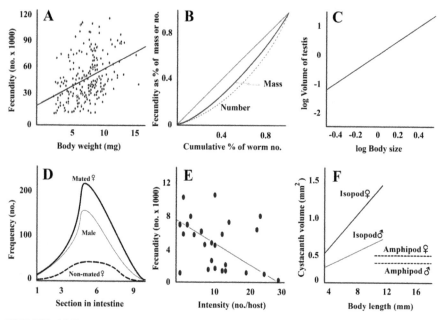

FIGURE 14.5

A. Fecundity as function of body weight of *Corynosoma cetaceum* (collected from fransiscana dolphin). B. Lorenz curves drawn from Gini coefficients for fecundity as mass or number of eggs in *C. cetaceum* (drawn from Aznar et al., 2018). C. Relation between body size and testis in *Acanthocephaloides propinquus* (modified and redrawn from Sasal et al., 2000). D. Distribution of *C. magdaleni* mated females, non-mated females and males in different sections of the intestine of seal *Phoca hispida saimensis* (modified and redrawn from Sinisalo et al., 2004). E. Fecundity as a function of parasite intensity in *A. propinquus* collected from *Gobius bucchichi* (drawn from Sasal et al., 2001). F. Cystacanth volume of *Acanthocephalus lucii* as a function of isopod size (intermediate host) (compiled and modified from Benesh and Valtonen, 2007).

and spawning in shallow waters. This seasonal periodicity coincides with the peak egg production of these acanthocephalans, when fish are in shallow waters, where appropriate IH is abundantly available (see Nickol, 2006).

The total number of oocytes contributing to fecundity is assured by waves of oogonial proliferation and subsequent oocyte recruitment (Pandian, 2013). Fecundity is a decisively important factor in recruitment and the population level (Pandian, 2018). For example, LF of *Corynosoma cetaceum* in *Pontoporia blainvillei* linearly and positively increases and growth is modest but highly significant, irrespective of the wide scattering of the values. A reason adduced for the scattering by Aznar et al. (2018) is that the Lorenz curves drawn on the basis of < 0.3 Gini coefficient indicates that a few dominant females produce far more eggs than the others at comparable body size. Further, the concavity is deeper for fecundity for the number rather than for the mass (Fig. 14.5B). Notably, the testis volume of *Acanthocephaloides propinquus* grows with increasing body size of *Gobius buchichii* (Fig. 14.5C) and suggests that sperm availability may not limit fertilization in larger worms. To establish themselves at an appropriate site providing nutrients and neutral pH, many acanthocephalans select the mid-intestine. In the seal *Phoca hispida saimensis*, *Corenosoma magdaleni* congregate in the seal's mid-intestinal section (Fig. 14.5D). Interestingly, this section provides them an ideal microhabitat for mating also. An estimate indicates the presence of ~ 80–90% mated females of *C. magdaleni* in the mid-intestine of the seal. With increasing intensity, *A. propinquus* in *G. buchichii*, LF is decreased (Fig. 14.5E) indicating the intense competition for food and space. Despite these, the cystacanth volume in *Acanthocephalus lucii* grows with increasing body size of the dorso-ventrally flattened almost oval shaped isopods (Fig. 14.5F). Notably, the trend for the female is at a higher level than that for the male, suggesting that the females survive longer and better than the males. However, the difference in the levels of the trend is narrowed for the worm in the laterally compressed almost cylindrical bodied amphipods. The limited space available per unit area in the cylindrical amphipods seems not to facilitate hosting more and more cystacanths with increasing body size. Remarkably, the sex-dependent space available is relatively larger in isopods of < 11 mm size but only beyond 11 mm size in amphipods. Notably, the cystacanth hosting isopod body size is limited to 12 mm but that of amphipod commences at 10 mm and terminates at 19 mm.

Part D
Schizocoelomata: Hemocoelomata

These include the most speciose two major phyla Mollusca and Arthropoda as well as the less speciose (total: 1,760 species) minor phyla from 144 speciose Pentastomida to 1,047 speciose Tardigrada. Of them, Tardigrada, Onychophora and Pentastomida are metamerically segmented hemocoelomates, which molt periodically or occasionally (e.g. Pentastomida). The other two phyla are true coelomates with (Echiura) or without the (Sipuncula) circulatory system. They are not metamerically segmented and are not known to molt. But they are capable of regeneration (Echiura), and regeneration and clonal multiplication is restricted to a single sipunculan species. The pentastomids are endoparasites but have not let any of their organs or tissue types to degenerate.

15

Priapulida

Introduction

The Priapulida are eutelic (Blackwelder and Garoian, 1986), ecdysic marine benthic worms that are characterized by radial cleavage (Budd, 2001) and deuterostomic development (e.g. *Priapulus caudatus*, Martin-Durain et al., 2012). Although often considered as pseudocoelomate (e.g. Hyman, 1951b), they have a true coelomic cavity lined by mesoderm (Land, 1970). With a radial (instead of spiral) cleavage and blastoporal anus (instead of a mouth) for an ecdysic protostomids, priapulids represent a turning point in evolutionary history. Despite the small number of species, they have attracted a some attention by evolutionary biologists.

The priapulid body is cylindrical in shape, warty in appearance and superficially annulated. It is divisible into an anterior presoma and a large trunk with a pair of conspicuous warty caudal appendages in *Priapulus* (Fig. 15.1C) but wanting in *Halicryptus* (Fig. 15.1A). The fat anterior end bears a wide circular mouth armed with spines. Posterior to it, is the invaginable proboscis. The posterior end bears the anus and two separate urinogenital openings (Fig. 15.1D). Their body wall consists of a cuticle, which is molted during ecdysis to enable the growth (Ruppert, 1991), epidermis, dermis, circular and longitudinal muscle layers. The space between the body wall and intestine is occupied by a cavity lined by a syncytial membrane of doubtful peritoneal nature. In this cavity, a pair of urinogenital organs is located one on each side of the intestine (Fig. 15.1D). The system consists of the gonoducts bearing the gonads; the ducts are opened to the exterior by respective pores at the posterior end (Hyman, 1951b). The gonad originates from a group of epithelial cells of the anterior part of the primary duct adjacent to the mesentery.

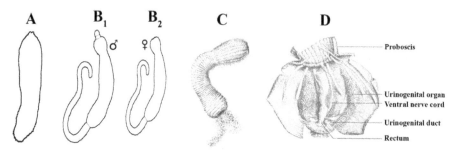

FIGURE 15.1

A. *Halicryptus spinulosus* (free hand drawing from Wennberg, 2008). B₁. Male and B₂. female of *Tubiluchus corallicola* (free hand drawings from Land, 1968). C. *Priapulus caudatus* and D. its anatomy (after Theel, 1906, 1911).

Taxonomy and Distribution

According to Land (1975), Priapulida comprises two orders. The Priapulidae with six species in four genera are macrobenthic predatory forms inhabiting muddy sediments in cold waters. *Tubiluchus corallicola* (Fig. 15.1B) is adapted to an intertidal life in sandy sediments in warm waters. The second order Seticoronaria include the tube-dwelling tentaculate species *Maccabeus tentaculatus* inhabits muddy bottoms of the Mediterranean Sea. The described number of priapulid species is increased from 9 (Land, 1975) to 19 (Wennberg, 2008). The rate of new species description is 0.3 species/y. Recorded reports for their vertical distribution are interesting: *M. tentaculatus* from 550 m in the Meditteranean Sea to 2,500 m for *M. cirratus* in the Indian Ocean and to 4,000 m for *Priapulus caudatus* and 8,000 m for *P. abyssorum* (see Wennberg, 2008). The priapulids are known for bipolar distribution: *Priapulus caudatus* and *Priapulopsis bicaudatus* in the Northern Hemisphere, and *Priapulopsis tuberculatospinosus* and *Priapulopsis australis* in the Southern Hemisphere (Land, 1975).

Reproductive Biology

The priapulids are gonochores. In them, sexual dimorphism is recognizable only in genus *Tubiluchus* (Fig. 15.1B). Sex ratio is slightly skewed in favor of females (e.g. 0.59 ♀ : 041 ♂, *T. troglodytes*, Todaro and Shirley, 2003). Fertilization is internal (Alberti and Storch, 1988), inclusive of the embryo-brooding *Meiopriapulus fijiensis* (Fig. 15.2G, Higgins and Storch, 1991). In

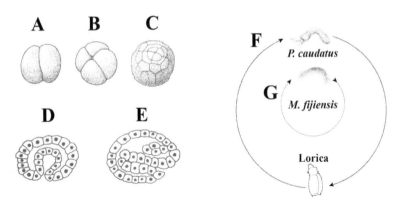

FIGURE 15.2

Early embryonic stages of *Priapulus caudatus*: A. 2-cell stage, B. 4-cell stage, C. 32-cell stage and D–E. gastrulation (from Land, 1953). F. Indirect life cycle involving lorica larval stage in *P. caudatus* (based on Wennberg, 2008). G. Direct life cycle in *Meiopriapulus fijiensis* (based on Higgins and Storch, 1991).

Priapulus caudatus, cleavage is commenced following fertilization. The first two cleavages are meriodional (Fig. 15.2A, B), the third is equatorial and the blastomeres are almost equal in size. In the 8th and 16th cell-stages, 2nd and 4th quartets are formed. The cleavage pattern is radial. At the 32-cell stage, a coeloblastula is formed; gastrulation is followed by invagination (Fig. 15.2D, E). Despite the availability of thousands of eggs, successful development up to hatching is limited to a few. In priapulids, the life cycle is indirect and the larva is named lorica (Fig. 15.2F). According to Janssen et al. (2009), the newly hatched lorica is still yolk-rich and lacks a mouth and pharyngeal teeth but is divisible into an introvert, neck and trunk. The initial lorica grows stepwise to a certain size prior to metamorphosis (Wennberg, 2008).

In deuterostomes, the blastopore forms the anus but its fate is variable protostomes. The orthologs of embryonic expression of the hindgut/posterior markers *brachyury* (*bra*) and *caudal* (*cdx*) are expressed in the blastopore and gut opening (amphistomy) of cnidarians, but in the blastopore, foregut and hindgut of spiralians and chaetognaths, and the blastopore and hindgut of arthropods. According to Martin-Duran et al. (2012), the invaginated endomesodermal cells in *Periapulus caudatus* develop into a small archenteron that opens to the exterior in the vegetal pole. The direct formation of the anus from the blastopore is supported by the embryonic expression of *bra* and *cdx* in the anus and hindgut. As indicated earlier, the formation of an anus from the blastopore in the protostomid ecdyzoan (see Fig. 1.3) *P. caudatus* marks the transition to deuterostomic development. The priapulids also represent unsegmented worms that are distantly related to arthropods. *Wnt* signaling is involved in segment border formation and regionalization of the segments, for example, in Annelida (Pandian, 2019). In *P. caudatus* and *H. spinulosus*, Hogvall et al. (2019) reported the presence of 12 *Wnt* genes and their expression in posteriorly located blastomere. Hence, *Wnt* genes may play an important role in posterior elongation of the trunk in priapulids.

16

Sipuncula

Introduction

The sipunculans are exclusively marine, unsegmented coelomatic worms characterized by anterior retractable introvert and posterior barrel-shaped or cylindrical trunk. A most conspicuous part of the worms is the introvert bearing a mass of 18–24 short tentacles. The worm size ranges from 2–3 mm in length (e.g. *Phascolion psammphilum*) to 50–60 cm (e.g. *Sipunculus indicus*) (Adrianov and Maiorova, 2010). Their body wall consists of a cuticle, epidermis, dermis, circular and longitudinal muscle layers, and peritoneum. Cuticular structures like spine, scales and holdfast (e.g. *Phascolion*) are produced by epidermis, which consists of abundant secretary glands. In some sipunculans, the dermis contains coelomic canals or spaces that communicate with the coelom through pores. The peritoneum consists of flattened epithelial cells provided with tufts of cilia, urns and chloragogue cells. In the coelomic fluid, hemerythrin is abundantly present. All sipunculids are provided with a single or a pair of metanephridium, an elongated organ(s) located in the anterior part of the trunk (Hyman, 1959). From the mouth, the coiled digestive tract of the worm passes to the posterior end of the body before twisting back around itself and ending at the anus on the dorsal side of the body (Fig. 16.1). The worms have a true coelom but do not have a vascular system. A separate cavity fills the hollow tentacles, which pass oxygen and nutrients from the tentacles to the coelom. The nervous system consists of a cerebral ganglion and a ventral nerve cord. When threatened, the worm retracts its body into a shape of a peanut. Hence, they are named peanut worms (Schulze, 2004).

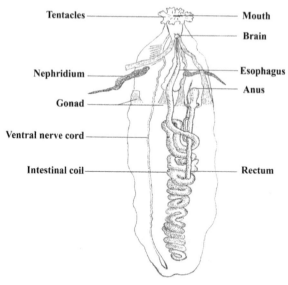

FIGURE 16.1

Internal anatomy of *Sipunculus nudus* (after Metalinikoff, 1900).

Taxonomy and Distribution

Typically, Sipuncula is divided into two classes, four orders, six families, 17 genera and 147 species (Schulze, 2004, Schulze et al., 2004). The class Sipunculidae is characterized by an array of tentacles surrounding the mouth (Fig. 16.2C) and the other class Phascolosomatidae, in which the tentacles are arranged in an arc around the nuchal organ (chemoreceptor on the tip of the introvert, Fig. 16.2D). These benthic worms have colonized intertidal to abyssal habitats from polar, temperate and tropical habitats (Boyle and Rice, 2014). They have been recorded up to the depths of 500 m in the North Pacific (Maiorova and Adrianov, 2017), 520 m (e.g. *Phascolosoma turnerae*, Rice et al., 2012) and up to 6,850 m (Schulze, 2004). Some of them occur in abundance, for example, 2,000/m² (e.g. *Themiste lageniformis*, Schulze, 2004) and 8,000/m² in the soft bottom (Adrainov and Maiorova, 2010).

Reproduction and Development

The sipunculids are gonochores with a single exception of the hermaphroditic *Nephasaoma minutum* (Schulze, 2004). They reproduce sexually, but *Themiste lageniformis* alone is known to reproduce parthenogenically (Pilger, 1987) and

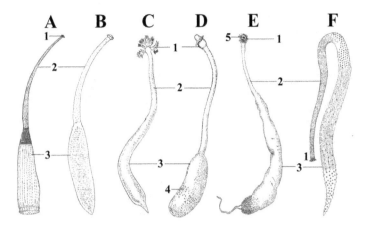

FIGURE 16.2

Some sipunculids: A. *Aspidosiphon speciosus*, B. *Golfingia minuta*, C. *Dendrostomum*, D. *Phascolion strombi* (after Thiel, 1875), E. *G. flagrifera* (after Selenka, 1995), F. *Phascolosoma nigrescens* (after Ward, 1891). 1 = tentacle, 2 = introvert, 3 = trunk, 4 = holdfast, 5 = mouth.

Aspidosiphon brocki alone clonally (Rice, 1970). In general, sex ratio is skewed in favor of females in some sipunculids (e.g. *Golfingia minuta*, *Phascolosoma agassizii*, *T. petricola*). In *T. lageniformis*, the male ratio ranges from 0.02 in Indian coast (Awati and Pradhan, 1936) to 0.03–0.04 and to 0.00 in different locations of the Florida coast (Pilger, 1987). *T. lageniformis* is a facultative parthenogenic. Pilger assumed that the males appear from unfertilized eggs that are developed spontaneously after spawning during autumn.

Reproductive system: In most sipunculans, the gonad extends as a narrow digitate band of tissue along the base of the two ventral retractor muscles, extending from or near the lateral edge of one muscle, under the ventral nerve cord to the edge of the muscle (Rice, 1975, Fig. 16.1). The gametes are shed into the coelom at an immature stage and undergo vitellogenesis and maturation, while floating in the coelomic fluid (Hyman, 1959). For more details on vitellogenesis and maturation, Rice (1975) may be consulted. The mature gametes are spawned through the nephridia, which act as gonoducts. Figure 16.3 shows gametes of some sipunculans. The eggs of sipunculans have a distinctive thick, perforated, multi-layered egg envelope (Fig. 16.3B). The spermatozoa may have a thin long flagellum, as in *Phascolosoma agassizii* (Fig. 16.3E) or short and stout flagellum, as in *Themiste pyroides* (Fig. 16.3C). Incidence of spermatozeugma (e.g. *Thysanocardia nigra*, Fig. 16.3D) is not uncommon. Six to seven months are required to mature in the coelom of *Phacolosoma arcuatum* and *Themiste lageniformis* (Cutler, 1994). With a direct development in *Golfingia minuta*, the period required for oocyte maturation

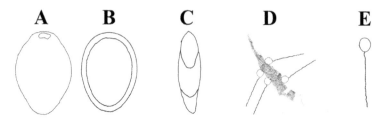

FIGURE 16.3

Sipunculan eggs: A. *Phascolosoma agassizii* (free hand drawing from Adrianov et al. (2016), B. *Phascolion cryptus* (from Rice, 1975), C. *Themiste pyroides* spermatozoan, D. spermatozeugma of *Thysanocardia nigra*, E. spermatozoa of *P. agassizii* (from Adrianov and Maiorova, 2010) (all are free hand drawings).

is even longer (10 months, Gibbs, 1975). As a consequence, a short or longer periodicity for spawning becomes obligatory.

From data summarized by Rice (1975), the spawning periodicity is plotted against latitude in Fig. 16.4. In general, the periodicity tends to coincide with summer months and corresponding temperatures at different latitudes, i.e. it is synchronized with respective to peak temperature during summer months, including *Phascolosoma arcuatum* during austral summer at 22°S in Queensland, Australia. It is sharp and restricted to June in *Paraspidisiphon steenstrupi* at 15°N. However, it is prolonged to five months (July to November) in *Phascolion cryptus* at 27°N and to seven months (March to August) in *Themiste pyroides* at 50°N. There are others, in which spawning occurs twice a year once in July at Panama (8°N) but from October to December in Venenzuela at 8°S for *Paraspidisiphon fischeri*. In *Phascolosoma perlucens*, it is first in April in Pueter Rico at 7°N and the second from October to February in Florida at 27°N. Hence, other than temperature, factors like food availability and period required for oocyte maturation may also regulate the time and duration of spawning periodicity.

Cleavage: According to Boyle and Rice (2014), cleavage is unequal, holoblastic and spiral. The first cleavage divides the egg into 2-blastomeres of unequal size, the large CD and smaller AB blastomeres. The second results in relatively large D blastomere and equal sized A, B, C blastomeres. After the third divisions, smaller 1A, 1B and 1C micromeres, that are smaller (lecithotrophic) than or similar in size (planktotrophic) to their respective 1a, 1b and 1c micromeres are produced. Gastrulation occurs by epiboly in species with lecithotrophic development but by invagination and/or epibody in species with planktotrophic development. The relatively large 1st quartet micromeres contribute to prototroch cells, which are considered to play an important nutritive role. The 4d blastomere divides to produce a pair of teloblasts that give rise to the mesodermal bands. As no cell-lineage experimental study has been made, direct cell-fate characterizations in

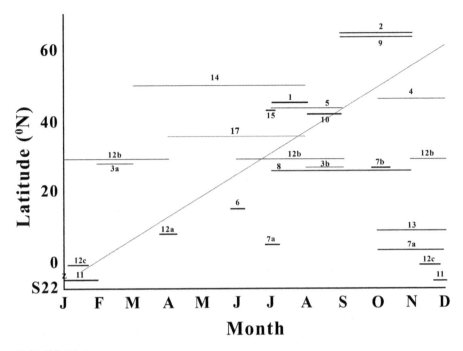

FIGURE 16.4

Spawning periodicity in some sipunculans as a function of latitude. Drawn from data summarized by Rice (1975). 1. *Golfingia elongata*, 2. *G. minuta*, 3a, b. *G. pellucida*, 4. *G. pugettensis*. 5. *G. vulgaris* (a) Panama, (b) Florida, (C), Venezuela, 6. *Parasiplophoron steenstrupi*, 7. *P. fischeri*, 8. *Phascolion cryptus*, 9. *Ph. strombi*, 10. *Phascolopsis gouldi*, 11. *Phascolosoma arcuatum*, 12. *Pha. perlucens* (a) Pueter Rico, (b) Florida, (c) Venenzuela, 13. *Themiste alutacea*, 14. *T. pyroides*, 15. *Sipunculus*, 16. *Siphonosoma cumanense*, 17. *Phascolosoma turnerae* (Rice et al., 2012).

embryos or later stages are not available. However, Rice (1975) and Adrianov et al. (2011) are referred for more details.

Larvae: In most sipunculans, the gastrula gives rise to a characteristic ovoid or top-shaped trochophore, which consists of pre-trochal and post-trochal hemispheres separated by a wide equatorial band of prototroch. The cells in the prototroch are ciliated and used for locomotion. The egg envelope persists through the trochophore stage functioning as an embryonic or larval covering. At metamorphosis, it may be either shed or transformed into a cuticle of the succeeding teleplanic pelagosphere larva. A fully developed trochophore may have a pharynx, stomach, gut and hindgut but the gut is non-functional. Hence, the sipunculan lecithotrophic trochophore is a non-feeding planktonic larval stage (Rice, 1989).

According to Rice (1989), the trochophore may metamorphose into pelagosphere larva in some sipunculans. The unique pelagosphere is characterized by a head, an expanded metatrochal area bearing a band of prominent metatrochal cilia and an elongate posterior trunk. There are two types of pelagospheres: (i) lecithotrophic and (ii) planktotrophic. Transition of the lecithotrophic to the juvenile stage is gradual and requires 3–4 weeks, as in *Themiste alutacea* and 7 weeks in *Golfingia pugettensis*. The feeding planktotrophic pelagosphere stay in the pelagic realm for periods longer than a month. Rarely, the duration of planktotrophic pelagosphere ranges from ~ 7 months to > 1 year in the wood-dwelling deep sea sipunculan *Phascolosoma turnerae*. Incidentally, the life span of *P. turnerae* is estimated at > 20 years. After a prolonged period, the planktotrophic pelagosphere undergoes a relatively rapid metamorphosis lasting for two to three days (Rice et al., 2012). With Generation Time (GT) as seven months and Life Span (LS) of 20 years, the GT as percentage of LS amounts to 2.9% in *P. turnerae* with long living pelagosphere. In the patterns 2 and 3 with shorter duration of larval stage, GT may amount < 2% of LS.

For sipunculans, no modern cell lineage or embryonic fate-map is available; genomic or transcriptomic data are also not available. The publication by Boyle and Seaver (2010) marks the first contribution to show the expression patterns of *foxA* and *GATA* transcription factor gene in *Themiste lageniformis*. Subsequent series of publication have traced the sequence of the expression pattern along the digestive tract in the directly developing sipunculan *Phascolion cryptus*. Five genes *foxA*, *xlox*, *cdx* (*caudal*), *bra* (*brachiury*) and *Wnt* are expressed in this order of sequence in the digestive system from ventral-anterior to dorsal-posterior direction, following the U-shaped configuration of the gut morphology (for figures, see Boyle and Rice, 2014).

Life Cycle

For 147 described sipunculan species, the life cycle is described for 22 species or 15% of sipunculans. This value may be compared with 3% for the free-living polychaetes (Pandian, 2019) and 3.6% for the parasitic digeneans (Pandian, 2020). Based on the egg size, the number of larval stage and lechithotrophic or planktotrophic pelagosphere larva, the sipunculan life cycle is recognized in four patterns. Accordingly, Pattern 1 includes the direct lecithotrophic development involving no larval stage in 3 species (Fig. 16.5A). Pattern 2 passes through an indirect development involving one pelagic lecithotrophic trochophore in 2 species (Fig. 16.5B). Pattern 3 involves indirect development with two (trochophore + pelagosphere) pelagic lecithotrophic larval stages in 6 species (Fig. 16.5C). Pattern 4 also involves indirect development with pelagic lecithotrophic trochophore and a long-living planktotrophic pelagosphere in 11 species (Fig. 16.5D). Figure 16.6 shows that the egg size ranges from 136 μm to 260 μm in Pattern 1. But the range is increasingly limited between 125 μm and 165 μm in Patterns 2 and 3, and between

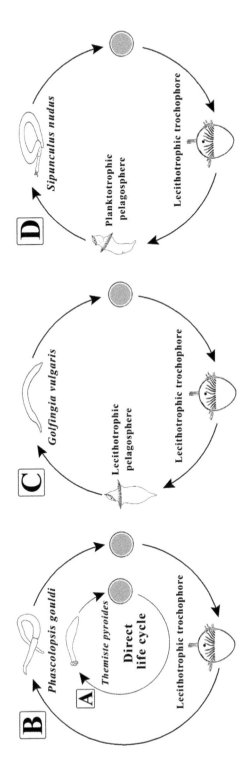

FIGURE 16.5

A. Direct life cycle in *Themiste pyroides*. B. Indirect life cycle involving lecithotrophic trochophore larva in *Phascolopsis gouldi*. C. Indirect life cycle involving lecithotrophic trochophore and pelagosphere larvae in *Golfingia vulgaris* and D. Indirect life cycle involving lecithotrophic trochophore and planktotrophic pelagosphere larvae in *Sipunculus nudus* (drawn on the basis of details from Boyle and Rice, 2014).

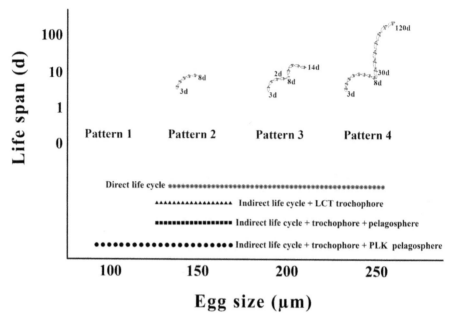

FIGURE 16.6

Egg size and life span of trochophore and pelagosphere in patterns 1, 2, 3 and 4 (drawn using data assembled in Table 16.1).

96 μm and 165 μm in Pattern 4. They stay in the pelagic realm for dispersal which is limited between 2–3 days and 8 days for the trochophore in Pattern 2 and 3. With inclusion of lecithotrophic pelagosphere, the stay is extended to ~ 11 days. In Pattern 4 too, the duration is limited between 2 and 9 days for the trochophore. But it is extended to the period of 30 days to 210 days for the planktotrophic pelagosphere and thereby provides enormous scope for growth and dispersal. For example, the wood-dwelling deep sea (366–1,184 m depth) *Phascolosoma turnerae* grows to a size of 350 mg (Rice et al., 2012) and may require 5 days to return to the benthics at the sinking rate 250 m per day (Pandian, 2020). Rice (1981) elaborated the advantages and problems encountered by sipunculan larvae adrift.

　　Of 22 sipunculan species, for which life cycle is known, 14, 9, 27 and 50% of them belong to Pattern 1, 2, 3 and 4, respectively (Table 16.1). In the primitive benthic sipunculans, pelagic development may have facilitated dispersal of the larvae. In them, life cycles have evolved in two opposite directions: One involved into the direction toward lecithotrophy with large but smaller number of egg/embryo along with reduction in duration of pelagic larval stage and direct development. The other involved the direction toward reduction in yolk, a large number of smaller egg and teleplanic planktotrophic larva that stays in the pelagic realm for a longer duration (see also Boyle

TABLE 16.1

Patterns and taxonomic distribution of development in sipunculans (compiled from Adrianov et al., 2011, Cutler, 1994, Johnson, 2001, Rice, 1975, 1981, Rice et al., 2012). PLK = planktotrophic, LCT = lecithotrophic

Pattern 1	Pattern 2	Pattern 3	Pattern 4
Egg size: 136 μm in *Phascolion cryptus* to ~ 260 μm in *Golfingia minuta*	Egg size: 125 μm in *Phascolion strombi* to 165 μm in *Phascolopsis gouldi*	Egg size: 125 μm in *Golfingia elongata* to 165 μm in *G. vulgaris*	Egg size: 96 μm in *Phascolosoma turnerae* to 165 μm in *Golfingia pelluscida*
Direct development. Embryo develops gradually into a juvenile passing through pelagic stage.	**Indirect development.** A pelagic trochopore, which transforms into a vermiform stage and finally into a juvenile	**Indirect development.** A pelagic LCT trochophore + LCT pelagosphore larvae	**Indirect development.** A pelagic LET trochophore + PLK pelagosphera larvae
Order: Golfingiiformes Family: Golfingiidae *Nephasoma minutum* Family: Phascolionidae *Phascolion cryptus* Family: Themistidae *Themiste pyroides*	**Order: Golfingiiformes** Family: Golfingiidae *Phascolion strombi* **Order: Sipunculiformes** Family: Sipunculidae *Phascolopsis gouldi*	**Order: Golfingiiformes** Family: Golfingiidae *Golfingia elongata* *G. pugettensis* *G. vulgaris* *Thysanocardia nigra* Family: Themistidae *Themiste alutacae* *T. lageniformis*	**Order: Golfingiiformes** Family: Golfingiidae *Apiansoma misakianum* *Golfingia pellucida* Family: Phascolosomatidae *Antillesoma antillarum* *Phascolosoma agassizii* *P. perlucens, P. turnerae, P. varians* **Order: Sipunculiformes** Family: Sipunculidae *Sipunculus nudus* *Siphonosoma cumanense* *S. polymyotus* **Order: Aspidosiphoniformes** Family: Aspidosiphonidae *Paraspidosiphon fischeri, Aspidosiphon parvulus*
	Trochophore: 3–8 days	Trochophore: 2–8 days **Pelagosphera:** **LCT: 2–14 d** **PLK: 0 d**	Trochophore: 2–9 days **Pelagosphera:** **LCT: 0 d** **PLK: 30 d, rarely 210 days**

and Rice, 2014). The latter seems to be the preferred direction, as 50% of the sipunculans pass through the planktotrophic pelagosphere larval stage.

Regeneration and Cloning

Between the 1880s and 1940s, when the presence of stem cells and their potency were unknown, experimental studies were made to understand the regenerative potency of the introvert, comprising mostly tissues of ecto- and meso-dermal origin and trunk consisting of ecto-, meso- and endo-dermal origin (see Fig. 16.1). In *Themiste zostericola*, the tentacles were regenerated, when the stumps were retained but not, when they were severed at the very base (Peebles and Fox, 1933). Clearly, the stumps retain the stem cells responsible for tentacular regeneration. Schleip (1934) found that (i) some regenerative cells appeared from the coelomocytes, which differentiated into mesodermal tissues and (ii) certain other cells (neoblasts?) arose from the ventral nerve cord and formed the epidermal and nervous tissues. From his studies on seven sipunculid species, Wegener (1938) reported a striking resemblance in the cytological characteristics of the migrating regenerative cells originating from the ventral nerve cord. In *Golfingia vulgaris* and *Aspidosiphon mulleri*, Bulow (1883) removed 3–7 mm of the extended anterior introvert including tentacles, brain (stomodeal part?) of the esophagus, contractile vessels, nerve cord and retractor muscles. After amputation, the wound was closed by the adjoining circular musculature of the body wall and a functional introvert was completely regenerated within 3–5 weeks. Clearly, the potency to regenerate mesodermal tissues and epidermal and nervous tissues is vested with the coelomocytes and nerve cells originating from the ventral nerve cord, respectively. In *Phascolion strombi*, a deep posterior incision led to the extrusion of the coiled intestine and subsequent death of the worm. Hence, the stem cells in the trunk do not have the potency to differentiate endodermal tissues (Schleip, 1935). When the incision was, however, limited to the body wall musculature of *G. minuta*, the wound was closed by contraction of the adjoining body musculature and sealed by the aggregated mass of amoebocytes, believed to originate from the mesodermal tissues. Cells migrated from the ventral nerve cord formed the ectodermal tissues of the regenerate. Therefore, the ectodermal and nervous tissues as well as mesodermal tissues can alone be regenerated but not the endodermal tissues in the sipunculan trunk.

It is in this context, a single reliable report by Rice (1970) on clonal reproduction in *Aspidosiphon brocki* becomes relevant. Firstly, she found that the clonal multiplication was limited to 15% of the field collected worms only. Evidently, 85% of the field collected or laboratory reared worms reproduce sexually alone. Secondly, the constriction, the induction point of clonal division was not initiated in any of the laboratory reared worms. But only the division was completed in the laboratory. Thirdly, the presence of gonadal tissue at the base of the ventral retractor muscles indicates their potential for sexual reproduction, albeit mature gametes have not been deducted in the coelom. Hence, it is not known whether or not the 15% of *A. brocki* is capable of sexual reproduction. Fourthly, her observation is thus far not reliably demonstrated in any other sipunculid species.

17

Echiura

Introduction

The 230 speciose (Zhang, 2011) Echiura are exclusively marine, cylindrical, soft-bodied coelomate spoon worms. Their body is divisible into a long (1 m long in *Bonellia*, 1.4 m long in *Ikeda taenoides*), scoop-shaped, non-retractable proboscis and trunk (Fig. 17.1). They have secondarily lost segmentation (Struck et al., 2007, Bourlat et al., 2008). Their muscular body surrounds a large coelom, extended into the proboscis but separated by a septum from the main body cavity in the trunk. The mouth opens into a long, highly coiled intestine, extended up to the posterior end of the trunk and terminates into a rectum, which opens to the exterior through an anus. A pair of simple or branched diverticula is connected to the rectum (Fig. 17.2). The diverticula are lined with numerous minute ciliated funnels presumed to be excretory in function. However, there are one to 100 metanephridia for excretion of nitrogenous waste, which typically open near the anterior end of the trunk. The worms do not have a discrete respiratory system; beside the body wall, cloaca (e.g. *Urechis*) absorbs oxygen. In most spoon worms, a distinct vascular system is present; the blood is colorless, albeit some hemoglobin containing cells are present in the coelomic fluid of the main body cavity. The

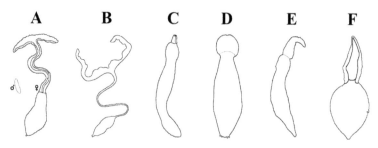

FIGURE 17.1

Some echiurans: A. *Bonellia*, B. *Ikeda taenioides*, C. *Echiurus*, D. *Urechis*, E. *Thalassema thalassemum*, F. *Listrolobus* (all are free hand drawings).

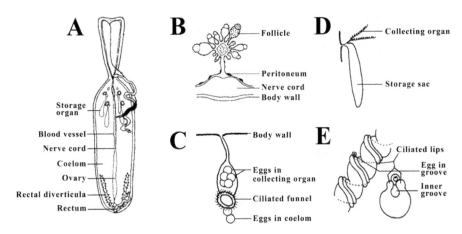

FIGURE 17.2

A. Anatomy of an echiuran (after Delage and Herouard, 1897). B. Cross section through *Bonellia* ovary (after Spengel, 1879). C. Simple collection and storage organ in *Ikeda*, D. complicated collecting and storage organ in *Urechis coupo* and E. details of the same (C–E are free hand drawings).

nervous system consists of a 'brain' located near the base of the proboscis and a ventral nerve cord running through the length of the trunk. The body size of spoon worms ranges from 2 mm in *Lissomyenema* to 2 m in *I. taenoides*. Their color is a dull gray or brown but a few species are brightly colored, for example, the translucent green *Listrolobus pelodes* (Walls, 1982). In South Korea, the spoon worms are eaten.

Taxonomy and Distribution

The 230 echiuran species are classified into two orders: (1) Bonellida comprising two families (i) Bonelliidae and (ii) Ikedidae, and (2) Echiurida consisting three families (iii) Echiuridae, (iv) Thalassematidae and (v) Urechidae (Fig. 17.1). Transcriptome analysis has revealed segment formation and secondary loss of it in *Urechis unicinctus* (Hou et al., 2019). The echiurans are strictly marine inhabitants and are found all over the world (Gould-Somera, 1975). The majority of echiurans live in soft sediments in shallow waters but some are found in rock crevices. They congregate in sediments with high concentration of organic matter; for example, the density of *Listrolobus pelodes* may reach 1,500/m^2 in the sediments (Stull et al., 1986). More than half the 70 species in Bonelliidae live below 3,000 m depth (Ruppert et al., 2004). Dawydoff (1959) reported the collection of some echiurans from a depth of 9,000 m. As sediment-feeders, echiurans play a major role in recycling nutrients in benthic communities.

Reproductive Biology

Echiurans are gonochores and reproduce sexually, except in *Urechis*, which may reproduce parthenogenically (Gould and Stephano, 1996). No hermaphroditism and clonal multiplication is thus far reported for any echiuran species. In them, sexes are not distinguishable externally, except by their gametes. Internally, the number of coelomoducts can be used to identify sexes in some thalassematids; in *Thalassema gogoshimense*, males with 32 pairs of coelomoducts can be distinguished from females with only 3 pairs (Gould-Somera, 1975). The bonellids, however, exhibit the most extreme sexual size dimorphism in the animal kingdom. The pygmy males of *Bonellia viridis* measure only 1 to 3 mm in length, in comparison to 1 m length in females.

Reproductive system: The echiurans seem to have adopted a different type of reproductive system. The system has three major components. (1) An unpaired gonad is located in the ventral mesentery just above the nerve cord (Fig. 17.2A). In *Urechis* spp, the gonad proper has never been found. (2) In both sexes, gametogenic cells are shed into the coelom, a major site of gametogenesis, in which vitellogenesis and maturation occur (Fig. 17.2B), as in sipunculans. (3) There are ducts to collect the mature gametes from the coelom and store them in storage organs until spawning. In the simple one, funnels with ciliated mouth collect the mature gametes, as in *Ikeda* (Fig. 17.2C). With lack of a closed circulatory system and coelom filled with blood cells, *Urechis* spp have developed a more complicated spiral collection system and storage organ (Fig. 17.2E). The collection process is not mechanical but involves selection at the ciliated lip. In *Urechis*, for example, small particles like sea urchin sperm are collected but among larger bodies, only ripe *Urechis* eggs are chosen. In some of them, up to 30 ml of ripe eggs and mature sperm can be stored (Fig. 17.2D).

Gametogenesis: Thanks to N.K. Das and his team, the process of spermatogenesis is known in *Urechis caupo*. At ~ 13°C, the process is commenced with premeiotic proliferation in spermatogonia in the coelom and is completed within 10 days. Figure 17.3A shows the correlated synthesis of protein, RNA and DNA during spermatogenesis. The oocytes (7–25 µm) are held by the ovary until fairly late in oogenesis. They are released, accompanied by a cap of accessory or peripheral cells in *Ikedosoma* and *Bonellia* (Fig. 17.3B). In the coelom, oocytes of different sizes ranging from 15–20 µm to 106–110 µm are present, i.e. asynchronously developing eggs are present almost throughout the year. The fully mature egg size ranges from 60 µm in *Echiurus abyssalis* to ~ 120 µm in *Urechis* spp, 190 µm in *Ikeda gogoshimense* and 400 µm in *Bonellia viridis* (see Gould-Somera, 1975).

Fertilization is external, except in Bonelliidae. However, MacGinitie (1935) injected the storage sac oocytes into a male coelom and observed

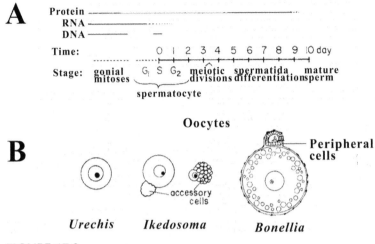

FIGURE 17.3

A. Spermatogenesis as function of time in *Urechis caupo* (free hand drawing from Das, 1968). B. Eggs of some echiurans (free hand drawings from Gould-Somero, 1975).

the development through metamorphosis. The very small (20 µm) oocytes of *Urechis* bind and let in sperm penetration but respond to Na⁺ dependent fertilization. However, they do not initiate meiosis and cleavage; hence, the fertilization process has inspired many studies. Investigating the molecular mechanism involved in fertilization, Gould and Stephano (1996) reported the following events in relation to time since insemination: 4 minutes: egg is rounded and the vitelline membrane begins to elevate; 7 minutes: germinal vesicle breakdown (GVBD) is begun; 22 minutes: nucleolus disappears; 35 minutes: first polar body is formed; 45 minutes: second polar body is formed; 70 minutes: pronuclei are fused; 90 minutes: cleavage is commenced.

Attempts have also been made to activate echiuran eggs parthenogenically. The transfer of the 3-minute old inseminated *Urechis* eggs to pH 7.1 stops the lifting of fertilization membrane, germinal vesicle breakdown and meiosis. However, it can be restored by transferring them to pH 8.2 (Lefevre, 1907). A reliable method for 100% activation in *Urechis caupo* is to expose their eggs to 0.1% trypsin in sea water. Yet, the eggs do not cleave, inspite of polar bodies being formed (Paul, 1970). On exposure to ammonical sea water, the activated *Urechis* eggs release polar bodies, but fail to cleave due to the loss of centriole (Tyler and Bauer, 1937). On the other hand, exposure to inorganic or organic acid in diluted sea water activates 40–60% of *Thalassema mellita* eggs but only very few trochophores are developed.

Cleavage and Larval Settlement

Echiurans are characterized by spiral cleavage and planktotrophic trochophore larva. In them, two patterns of embryonic and larval development are recognized: Pattern 1: The *Urechis* pattern, in which the embryo passes through coeloblastula, gastrulation by invagination and 2–3 months long planktotrophic trochophore (Fig. 17.4A) and Pattern 2: The *Bonellia* pattern with embryo passing through stereoblastula, gastrulation by epiboly and 7–10 days long lecithotrophic trochophore (Fig. 17.4B). In *Ikedosoma*, *Thelassema* and *Urechis*, the first few cleavages are nearly equal (Fig. 17.4A). In them, a coeloblastula hatches by pushing the cilia out through the vitelline membrane (Gould-Somero, 1975). At the 148-cell stage, an archenteron is formed, following gastrulation by invagination (Davis, 1989). A free swimming planktotrophic trochophore is released within 22 hours after fertilization in *T. mellita* but 40 hours in *Urechis*. Their larvae are adorned with the largest and most powerful prototrochs to enable their swimming activity. There are other ciliary bands to enable their feeding during their stay in the pelagic realm for two to three months. In contrast, an upper 4 micromeres and lower 4 macromeres are produced at the 8-cell stage in *Bonellia*. The large yolk droplets segregate into the macromeres (Fig. 17.4B). The embryo develops an unciliated stereoblastula. Similarly, gastrulation is accomplished by an epiboly instead of invagination, as in *Urechis* type. Its embryo is hatched only after gastrulation. The lecithotrophic trochophore is

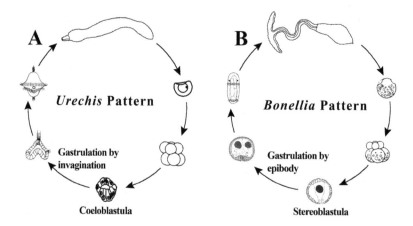

FIGURE 17.4

In the (A) *Urechis* pattern the life cycle involves 75-day long embryonic and planktotrophic larval stages but (B) in the *Bonellia* pattern, it involves 25-day short embryonic and lecithotrophic larval stages (free hand drawings from Gould-Somero, 1975, Davis, 1989 and others sources).

covered with a fine cilia and has two prominent bands of large cilia, two eye spots and so forth—all of them enable the larva to become motile and crawl around for a period of a week. Consequent to the differences in cleavage pattern and larval stage, there are distinct differences in their settlement pattern (Pilger, 1978). Firstly, the planktotrophic trochophores are negatively geotrophics. For example, *Echiurus abyssalis* larvae rise to the surface from the depths of 1,900 m. Their settlement is influenced by the nutrient property of the sediment; for example, *Urechis* larvae settle preferentially on sediments with high organic content (see Pilger, 1978).

Sex Differentiation

The spoon worm *Bonellia viridis* is a classical example for the renowned sexual size dimorphism with its bizarre form of sex differentiation. Not surprisingly, *B. viridis* has attracted the attention of biologists since 1868 by Kowalevski. According to Berec et al. (2005), *B. viridis* dwells in a burrow. A vast majority of its sexually undifferentiated larvae metamorphose into males, when exposed to females but differentiate into females, when developing in the absence of female(s). The free-living females have a cylindrical trunk of 8 cm length and an extendable prostomium to over 1.5 m. But, the dwarf (1–3 mm long) planarian-like testicular males enter and permanently inhabit the androecia, the modified nephridia of females. Up to 85 males may settle on the proboscis of a single female but only a maximum of four of these males are successful in settling in on androecium. The females spawn once a year producing an egg string with ~ 1,000 eggs. The string is kept in the female's burrow until the eggs are more or less simultaneously hatched into trochophore larvae. Following a period of 7 days lecithotrophic stay in and around the bottom, the trochophores begin to settle in vacant burrows, which become available through mortality of females of the previous generation. The dispersing ability of the trochophore is 8.4 m for 1% of the larvae and 10.4 m for the 0.1% of the larvae. The patchily available vacant burrows are quickly colonized by the competing simultaneously hatched larvae pending to settle soon. Hence, the settling larvae may have to decide whether to become males and compete for fertilization opportunity or to become females and compete for the limited number of suitable burrows.

At a time, when not much was known on biology and population dynamics, Bultzer (1931), Zurbuchen (1937) and Berec et al. (2005) undertook simple experiments to show that the larval settlement and its duration on the female proboscis induced sex (wrongly stated as) determination by authors including Pilger (1978) instead of differentiation. Bultzer showed that 86% larvae settled in clean water were differentiated into females. Contrastingly, 71% larvae that settled on female's proboscis were differentiated into males (Table 17.1). Zurbuchen demonstrated that the exposure duration of settling

TABLE 17.1

Sex differentiation in *Bonellia viridis* larvae exposed to different types of exposure and duration (compiled from Bultzer, 1931, Zurbuchen, 1937)

Exposure/Duration	♀ (%)	♂ (%)	Intersex (%)	Dead or Lost (%)
Type of exposure				
Clean seawater	86.0	1.4	0.3	5.6
♀ proboscis	20.2	70.7	0.5	8.5
Duration of exposure				
7–8 hours	15.3	61.0	23.7	-
13–33 hours	1.6	95.2	3.2	-

larvae on the female increased male differentiation from 61% with 7–8 hours exposure to 95% with 13–33 hours exposure. Incidentally, the prototroch and telotroch of the settling larvae on the female are lost within 13.5 hours, i.e. after 14 hours exposure, the larvae reached the point of no return, as indicated by the decrease in intersex from 24% in those exposed to 8 hours duration to 3% in those exposed to > 13 hours duration. On reaching the point of no return, the proboscis of the male larvae becomes virtually non-existent. The nerve ring becomes deflected posteriorly and so forth. Eventually, the metamorphosed male consists of a small ciliated body, seminal vesicle, a gut without mouth and anus, and a pair of metanephridia (see Pilger, 1978). Following Bultzer and Zurbuchen, a series of publication revealed that a thermostable aqueous extract of *B. viridis* intestine was also effective, when used at concentration of one part (by weight) of dried tissue to 6,000–9,000 parts of sea water. Bultzer's theory that proposed a diffusible substance inhibits the female differentiation or induces male differentiation, was supported by subsequent reports. One report indicated that the completeness of male differentiation is proportional to the duration of larval settlement on the female. For, the development of the seminal vesicle and production of viable sperm require longer (> 14 days) duration but the threshold for the pre-oral lobe reduction requires only a weak stimulation (see Pilger, 1978).

A second line of investigation into sex differentiation was pioneered by Herbst (1928). He demonstrated that by varying the concentration of certain elements in sea water, it is possible to masculanize 70–80% of the undifferentiated larvae, in the total absence of females. For example, increasing the concentration of copper or potassium, or decreasing that of magnesium or sulfate masculanizes 90% larvae. Even increasing acidity favored the male differentiation. A third line of investigation was that of Wilczynski (1968). Based on reactivity to a specific stain, he claimed that there are two distinct egg types, which distinguish the male from female tissues. Re-examining the controversy between Bultzer and Wilczynski, Leutert (1974), who repeated the Bultzer's experiment, showed that sex is determined genetically in

43–83% (an average of 63%) of larvae. This may be followed by a rigid and stable sex differentiation in some eggs but in the remaining phenotypes, the differentiation is labile. In this context, it is interesting to note that amputation studies have shown the existence of 67% genetic males and 33% phenotypic males in the not too distantly related annelid *Ophryotrocha labronica* (Pandian, 2019). As can be noted from the following account, the sex differentiation process in *B. viridis* is likely to be a labile process and is regulated by one or more overriding autosomal gene(s) induced by environmental factor(s).

At this point of time, it is necessary to elaborate the concept of (i) Genetic Sex Determination (GSD) and Environmental Sex Determination (ESD) as well as (ii) Genetic sex differentiation (Gsd) and Environmental sex differentiation (Esd). Sex determination and differentiation are diverse but related processes and the latter succeeds the former. Sex is determined by genes through sex chromosomes at fertilization but it is differentiated through genes controlling the endocrine pathway; this process may be stable, as in mammals and birds but labile, as in reptiles and fishes. In animals, sex is determined either by a gene harbored on male heterogametic chromosome (e.g. XX-XY, as in Nile tilapia, Pandian, 2017) or XX-XO, as in the nematode *Caenorhabditis elegans* or female heterogametic (e.g. ZZ-ZW), as in *Schistosomatium douthitti* (Pandian, 2020), or a set of sex chromosomes XX-X_{4-7}Y, as in some ostracods (Pandian, 2016).

Considering the sex differentiation process in some fish and ostrocods, this aspect may further be elaborated. Type 1 Genetic sex differentiation (Gsd): The process may solely depend on genes that are not amenable to induction by any environmental factor (e.g. the demersal fish *Hipploglossus hippoglossus*, Hughes et al., 2008). Type 2 Gsd + autosomal genes: In this type, exemplified by *Oreochromis niloticus*, male sex is determined by *Dmrt* harbored on a morphologically distinguishable Y chromosome. However, one or more overriding autosomal gene(s) derails the differentiation process in some individuals resulting in phenotypic female, while many individuals may realize the genetic sex differentiation. The overriding autosomal gene(s) may express inherently, as in *Puntius tetrazona* (Kirankumar and Pandian, 2003) or may be induced to express by one or other environmental factors like temperature, pH and hypoxia (Pandian, 2015). Type 3: In this pathway, sex is determined by a combination of many genes, each with small additive effects that are distributed in the number of X chromosomes, as in ostracods (Pandian, 2016). While passing through an extended Gsd labile pathway, the genetic sex may be altered to the opposite phenotypic sex by one or other sex autosomal genes, whose expression is under the control of environmental factors. Esd is widespread in fishes and reptiles and is likely in *Bonellia*. In contrast Gsd is stable and rigid in most echinoderms (Pandian, 2018), and molluscs (Pandian, 2017).

18

Tardigrada

Introduction

Typically, Tardigrada are barrel-shaped segmented microscopic (maximum body length of 1.2 mm, Nelson et al., 2010) pseudocoelomatic (rather hemocoelomatic ?) water bears. But their vestigial eucoelom is retained in the adult gonocoel (Pollock, 1975). Their bilaterally symmetrical body is divisible into a head and trunk with four segments, each bearing a pair of unsegmented lobopods, usually terminating in claws (Fig. 18.1). Their body is convex on the dorsal side but flattened on the ventral side; it is fully covered by a cuticle and displays limited metamerism (Pollock, 1975). The cuticle is molted periodically throughout the life (Gross et al., 2015). The morphology of the cuticle varies; in some, it is not chitinous (Pollock, 1975) but chitinous in others (Gross et al., 2015). Molting requires 5–10 days (Nelson et al., 2010). The ventrally located terminal mouth leads into a buccal bulb, followed by sucking muscular pharyngeal bulb and the gut terminating in

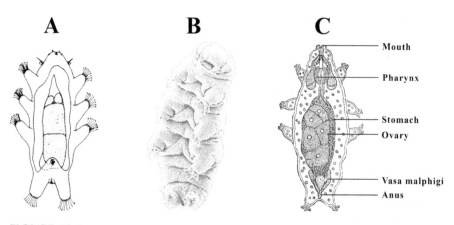

A	B	C

Mouth

Pharynx

Stomach

Ovary

Vasa malphigi

Anus

FIGURE 18.1

A. *Echiniscoides* sp (free hand drawing from Pollock, 1975). B. *Hypsilius dujardini* (from Rpgch Wikimedia Commons). C. *Macrobiotus* (from sysopje@yahoo.com).

a cloaca. Malphigian tubules are the main organs of excretory system. The nervous system consists of a dorsal brain, four-trunked ganglia which are interconnected by nerves. No respiratory or circulatory system is described (Nelson et al., 2010, Gross et al., 2015). Reared in a laboratory, the carnivorous *Paramacrobiotus ritchersi* survived up to 518 days and herbivorous *Diphascon* cf *scoticum* up to 203 days (Nelson et al., 2010).

Taxonomy and Distribution

The phylum Tardigrada comprises 2 + 1 classes, 4 orders, 21 families, 106 genera and 1,047 species (Nielson, 2013). Molecular studies with 80s rRNA have confirmed their phylogenetic proximity to Arthropoda and Onychophora, and the three are together named as the monophyletic taxon Panarthropoda. This morphology based classification and habitat distribution are described in Table 18.1. For 1,047 species, Nelson et al. (2010) provided details on habitat distribution for 160, 2 and 284 species as marine, freshwater and terrestrial inhabitants; the remaining 605 species are limno-terrestrial inhabitants. Hence, the Tardigrada are predominantly terrestrial inhabitants and are hydrophilous micrometazoans feeding on bacteria and the like.

TABLE 18.1

Classification of Tardigrada and their habitat distribution (based on Gross et al., 2015). T = terrestrial, FW = freshwater, M = marine

1. Class: Heterotardigrada	**3. Class: Eutardigrada**
Order: Arthrotardigrada M	Order: Apochela
(7 families, 39 genera, 149 species)	Family: Milnesliidae: FW + T (3 genera, 15 species)
Order: Echiniscoidea	Order: Parachela
Family: Echiniscoididae M	Family: Beornidae (1 fossil species)
(2 genera, 11 species)	Macrobiotidae: T + FW (13 genera, 229 species)
Carphaniidae: FW (1 genus, 1 species)	Murrayidae: T + FW (3 genera, 21 species)
Oreellidae: T (1 genus, 2 species)	Eohypsibiidae: T + FW (2 genera, 8 species)
Echiniscidae: T (12 genera, 242 species)	Hypsibiidae: T + FW (337 species)
2. Class: Mesotardigrada	Microhypsidiidae: T + FW (2 genera, 5 species)
Family: Themozodiidae: FW	Calohypsibiidae: T (5 genera, 21 species)
(1 genus, 1 species)	Necopinatidae: T (1 genus, 1 species)

Reproductive Biology

Most Tardigrada are gonochores (in heterotardigrades). But a few are thelytokic parthenogens or hermaphrodites (in eutardigrade families), especially in limno-terrestrial habitats. Table 18.2 provides a summary of reproductive modes in tardigrades. Sporadic incidence of hermaphroditism and/or parthenogenesis is not uncommon among limno-terrestrial eutardigrades. Hermaphroditism is known from Macrobiotidae, Eohypsibiidae and Calohypsibiidae, Hypsibiidae and a single necopinatid species. In others, cross fertilization is the rule. Ameiotic (apomictic) or meiotic (automictic) parthenogenesis occurs in some Macrobiotidae and Hypsibiidae. Males may be semelparous in some tardigrades milnesiids and hypsibiids. In others, iteroparity is common among females.

In tardigradees, the saccular gonad is suspended dorsally in the pseudocoelom by ligaments from its anterior end to insertions on the cuticle above the pharynx. They are developed from the fusion of one of the four pairs of somatic coelomic pouches during organogenesis. Hence, the gonocoel represents the vestige of the eucoelom retained in adults (Pollock, 1975). From the gonad, a pair of duct emerges and opens into the cloaca anterior to the anus. The gonopore is tubular in the male but radial in the female. In females, a pair of receptacles is present. The ovary contains at the maximum four or eight oocytes, which are irregularly distorted and are approximately equal in size and appearance. Eggs are laid only during molting. The flagellated sperm of Tardigrada ranges from 80–90 μm in length. Mating requires the proximity of the female and male during the female's molting period. Only in *Hypsilius nodusus*, is the internal fertilization described.

Cleavage and Development

In Tardigrada, the life cycle is direct. Cleavages are total and blastomeres are equal in size (Fig. 18.2). Up to a 16-cell stage, cleavages are synchronized. Subsequently, a coeloblastula with a small blastocoel is formed. A stereogastrula is developed through gastrulation by unipolar proliferation, wherein a few cells from the vegetal pole are released into the blastocoel and subsequently divided to form 50–60 primary endodermal cells. Amidst it, the archenteron is also developed. Of six enterocoelus pouches that are then formed, two are cephalic and four are somatic coelomic pouches. No secondary body cavity, i.e. deuterocoel is formed. The cephalic pouches give rise to the buccal apparatus; the anterior most three pairs of somatic pouches disaggregate to form endomesodermal structures like coelomocytes and muscle bands. The dorsal gonad is developed from the fourth pair of pouches, which remain attached to the archenteron (Fig. 18.1).

TABLE 18.2

Summary of reproductive modes in tardigrades (modified from Bertolani, 2001). Y = yes, present

Traits	Heterotardigrades				Non-marine Eutardigrades											
	Marine		Non-marine		Milnesiidae		Macrobiotidae		Eohypsibiidae		Calohypsibiidae		Hypsibiidae			Necopinatidae
	♂	♀	♂	♀	♂	♀	♂	♀	♂	♀	♂	♀	♂	♀	♀	♀
Fertilization																
Selfing	-	-	-	-	-	-	-	Y	-	Y	-	Y	-	-	Y	Y
Crossing	Y	-	Y	-	Y	-	-	Y	-	Y	-	Y	Y	-	-	-
Parthenogenesis																
Meiotic	-	-	-	-	-	-	Y	-	-	-	-	-	-	Y	-	-
Ameotic	-	-	-	-	-	-	Y	-	-	-	-	-	-	Y	-	-
Unidentified	-	-	-	Y	-	Y	-	-	Y	-	-	-	-	-	-	-
Semelparity	Y	-	-	-	Y	-	-	-	-	-	-	-	Y	-	-	-
Iteroparity	-	Y	-	Y	-	-	Y	Y	Y	Y	-	Y	Y	Y	Y	-

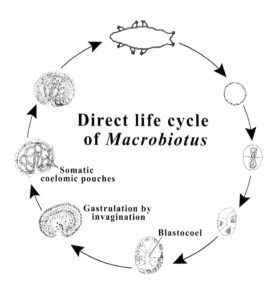

FIGURE 18.2

Direct life cycle of *Macrobiotus*. For completion, one stage from *Hypsilius* is included (compiled and modified from after Van Wenck, 1914, after Marcus, 1929).

Surviving Strategies

As the majority of tardigrades are terrestrial and/or limno-terrestrial inhabitants (living in water film on land), they are subjected to extreme vicissitudes in temperature and humidity. They have developed adaptive dormant strategies to survive these extremes. There are two forms of dormany, (i) cryptobiosis and (ii) encystment (Guidetti et al., 2006). The ability of tardigrades to undergo cryptobiosis is more widely known than their encystment behavior. In tardigrades, the cryptobiosis is exhibited in the following forms: (1a) Anhydrobiosis induced by loss of water or desiccation, (1b) Osmosis induced by loss of water due to higher external salt concentration (Bertolani et al., 2004). (2) Cryobiosis induced by freezing temperature and (3) Oxybiosis induced by hypoxia/anoxia. In tardigrades, the most common form of crptobiosis is anhydrobiosis, in which alone tuns are formed.

Anhydrobiosis: Alongside it, tardigrades have developed one set of a strategy to prevent cell destruction and the other to release energy to sustain the dormant period. Tardigrades have developed behavioral adaptations like the formation of tun and aggregation. Tun is a barrel-shaped, dry dormant tardigrada. To develop the tun, they retract the head, lobopods and posterior, and form a rounded body, and thereby reduce the surface area. The second behavioral adaptation is aggregation, which reduces the surface area of an individual tun. The imminent anhydrobiotic tardigrade stores nutrients to sustain itself during dormancy. In *Milnesium tardigradum, Paramacrobiotus tonellii* and *Macrobiotus sapiens,* Reuner et al. (2010) described that the free floating cells store the energy releasing glycogen and others. With the increasing storage cell size, greater is the probability of survival during anhydrobiosis for the median sized animals than that for the larger sized *Richtersius coronifer* (Fig. 18.3A). Not surprisingly, anhydrobiosis is restricted to tardigrades of < 1 mm in length (Watanabe, 2006).

To prevent radiation from causing ionization, tardigrades replaces water in the cell membrane with sugar. The sugar replacement prevents the rupture of the cell wall (Brave New Biosphere, 1999). They lose most of their free and bound water (> 95%) and reduce or suspend their metabolism. Their survival is correlated with the synthesis of cell protecting Heat Shock Protein (HSP) of several major chaperons families (e.g. *M. tardigradum,* Schokraie, 2011), trehalose and glycerol. In *R. coronifer,* trehalose accumulation reaches 0.1 to 2.3% of dry weight within 5–7 hours following desiccation. Of multiple properties of trehalose listed by Watanabe (2006), structuring activity of intracellular water with HSP and stabilization of dry membranes are relevant to desiccation. It modifies the physical properties of membrane phospholipids and stabilizes the dry membrane.

Anhydrobiosis, which stops the biological clock during dormancy, extends the life span of tardigrades. For example, it is 1–2 years in limnic

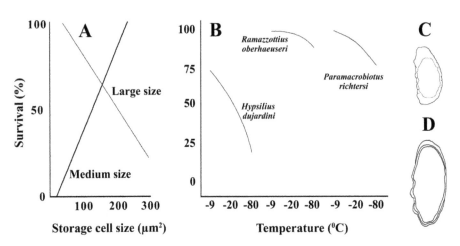

FIGURE 18.3

A. Probability of survival as a function of storage cell size in anhydrobiotic medium and large sized *Richtersius coronifer* (modified and redrawn from Jonsson and Rebecchi, 2002). B. Survival of bryophyte-dwelling *Hypsilius dujardini, Ramazzottius oberhaeuseri* and *Paramacrobiotus richtersi* subjected to sub-zero temperatures (redrawn from Bertolani et al., 2004). C. Type 1 cyst surrounded by a single cuticular layer and D. type 2 cyst by multiple cuticular layers (Sketches from Glime, 2017b).

Hypsilius and *Macrobiotus* but 4–12 years in terrestrial bryophyte-inhabiting species of the same genera, which often undergo anhydrobiosis. Some tardigrades are claimed with a century-long survival in an anhydrobiotic state. A herbarium specimen of a moss housed a tardigrade began cellular activity after a period of 120 years of being dry in the herbarium (Jonsson and Bertolani, 2001). Rebecchi et al. (2009b) tested this claim of longevity by allowing 5 tardigrade species to a dry ambient condition. Among them, 91% of *Ramazzottius oberhaeuseri* and 72% of *Echiniscus testudo* survived for 1,604 and 1,085 days, respectively. Guidetti and Jonsson (2002) found that of 63 samples from stored collections kept in anhydrobiotic state for 9–138 years, the eggs survived longer than the dormant tuns. The reasons for reducing longevity beyond a particular duration are the destruction of trehalose by continued high temperature (Rebecchi et al., 2009a) and structural changes in DNA (Newmann et al., 2009).

Cryobiosis: Tardigrades can survive brief exposure to $-273°C$ (Miller, 1997). The desiccated dormant state lets them survive pressure of 6,000 atmosphere (Seki and Toyoshima, 1998)—six times the pressure of the deeper parts of the ocean. Frozen tardigrdes can survive for several years (see Nelson et al., 2010). To survive low temperature, it is important to prevent freezing of the cellular water. The cryobiotic tardigrades accomplish it by production of ice-nucleating proteins that are located outside the cells. These proteins act like water magnets drawing water out of the cell and render the cell to become unfreezable. *R. coronifer* living amidst mosses overwinters in a frozen or dry state. On the other hand, *Bertolanius nebulosus* dwelling among moist mosses and algae overwinters as eggs. However, both of them can supercool $-7°C$. In them, crystallization and ice formation are prevented within a few minutes of nucleation (Westh et al., 1991). *R. oberhaeuseri* and *Paramacrbiotus richtersi* survive 80°C but that of *Hypsilius dujardini* is 75 and 30% at $-9°C$ and $-80°C$, respectively (Fig. 18.3B). Soil-dwelling tardigrades do not super cool but survive by dehydration.

Anoxybiosis results in failure of osmosis, which causes the entry of excess water into the cell. The animal becomes turbid and immobile. This strategy is used by tadigrades living in Antarctic lakes to overwinter in the anoxic deep water (see Nelson et al., 2010).

Encystment or cyst formation or diapause is another survival strategy adapted by freshwater, moss-dwelling and soil-living tardigrades. During encystment, new cuticular layers are synthesized. Two types of cuticular enclosures are recognized. In *Dactylobius parthenogenicus*, a new cuticle, similar in structure to the old one, is formed. In two phenotypes of *Bertolanius volubilis*, one with a single cuticular layer and the other with multiple cuticular layers are formed (Fig. 18.3C, D). The cysts can survive for months in nature without totally depleting nutrient resources (see Nelson, 2010). Some tardigrades can simultaneously produce both subitaneous (non-encysting) and encysting eggs. In *P. richtersi*, Altiero et al. (2006) found three types of hatching:

(i) subitaneous eggs hatching in 30–40 days and (ii) delayed hatching between 41–62 days; in (iii) diapausing (encysted) eggs, in which 30% of the eggs require a drying period followed by a rehydration period prior to hatching after 90 days but the remaining 67% eggs are not known to hatch. Altiero et al. (2006) traced the life history traits of two phenotypic clones in parthenogenic *Macrobiotus richtersi*; clone 1 is more fecund and fertile than those of clone 2 (Table 18.3). With a wide difference in the life span of these clones, *M. richtersi* has developed different survival strategies to inhabit water bodies, in which water may be available for a short or longer duration.

TABLE 18.3

Life history traits in two clones of *Macrobiotus richtersi* (condensed from Altiero et al., 2006)

Life History Trait	Clone 1	Clone 2
Life span	195	137
Age at I oviposition (d)	77	71
Oviposition (no./life span)	3.4	2.0
Batch fecundity (egg no.)	11	8
Life time fecundity (egg no.)	38	18
Hatching duration (d)	41	60
Hatching success (%)	83	51

19

Onychophora

Introduction

Onychophorans are terrestrial, segmented hemocoelomate velvet worms (Fig. 19.1A, B). Their body consists of a head, trunk and an anal cone. The three cephalic segments are modified into annulated, fleshy antennae, jaws and slime papillae, that eject a sticky secretion used to capture prey or as defense. In the trunk, each segment bears a pair of unjointed lobopods; each

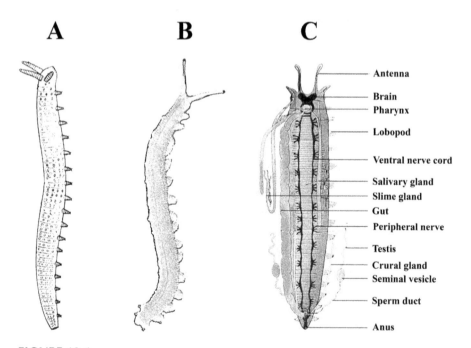

A   **B**   **C**

Antenna
Brain
Pharynx
Lobopod
Ventral nerve cord
Salivary gland
Slime gland
Gut
Peripheral nerve
Testis
Crural gland
Seminal vesicle
Sperm duct
Anus

FIGURE 19.1

A. *Peripatus* (from biologydiscussion.com). B. *Ophithopatus highveldi* (from Stellenbosch University). C. Anatomy of *Peripatus* (from biologydiscussion.com).

lobopod is equipped with a pair of scelertized claws. Their body is covered by a thin, flexible, permeable chitinous cuticle, which is periodically molted. Between the cuticle and muscular layer lays a thin epidermis; the latter consists of circular, diagonal and circular-muscular layers. Their coelom is partitioned into sinuses and the true coelom is restricted to gonadal cavities. Their respiratory system consists of numerous tracheas, which open to the exterior via specialized atria. A well-developed circulatory system is present; the tubular heart opens at both ends and is located in the pericardial sinus. The colorless blood enters the heart through ostia and leaves it anteriorly and flows throughout the hemocoel within the body sinuses. Hence, oxygen is delivered by trachea and nutrients by blood. Segmental nephridia serve as excretory organ, most of which open to the exterior near the base of each lobopod. Their bilobed 'brain' is connected by two widely separated nerve cords (Fig. 19.1C). The onychophoran nervous system lacks metameric ganglia. The body size ranges from 5 mm to 15 cm. The onychophoran nervous system lacks metameric ganglion. Their body size ranges from 5 mm to 15 cm and color can be blue, black, green or orange (Mayer et al., 2015, Wright, Animal Diversity Web).

Taxonomy and Distribution

The 200 speciose onychophorans are divided into two families: Peripatoidae and Peripatosidae. They are found in dark moist microhabitats and are distributed in mutually exclusive geographic regions. The former is circumtropical, found in Southeast Asia, equatorial Africa, and South and Central America, and the latter circumaustral in Chile, South Africa and Austrasia (Mayer et al., 2015). The peculiar mutually exclusive geographic distribution has imposed distinct differences in some life-history traits of onychophorans. The peripatids dwelling in relatively warmer regions live longer, delay sexual maturation and are more fecund than the periaptosids (Table 19.1).

Reproductive Biology

Onychophorans are gonochores, reproduce sexually and pass through a direct life cycle. In a few species, the female ratio is skewed; the reason for this skewing may be traced to birth in captured rearing, as in *Macroperipatus torquatus*. That the ratio in *Plicatoperipatus jamaicensis* embryo is 0.50 but 0.62 in the field indicates a greater mortality of males (Table 19.2). Females are mainly are larger than males (e.g. *Euperipatoides rowelli*, Sunnucks

TABLE 19.1

Differences in some life history traits between Peripatidae and Peripatosidae (summarized from Monge-Najera, 1994)

Traits	Peripatidae	Peripatopsidae
Maturation age (month)	30	15
Fecundity (no.)	23	9
Parental investment (young as % of mother)	6.9	4.6

TABLE 19.2

Female ratio in some onychophorans (modified from Monge-Najera, 1994)

Peripatidae	
Peripatus fuliformis	0.75
Oroperipatus eiseni	0.65
Macroperipatus torquatus[††]	0.65
M. trinidadensis	0.65
Plicatoperipatus jamaicensis[†]	0.62
Pe. acacioi	0.61
Epiperipatus biolleyi	0.54
*M. torquatus***[**]	0.53
M. trinidadensis	0.53
Pl. jamaicanensis[*]	0.50
Pe. heloisae	0.48
Peripalopsidae	
Paropisthopatus umbrinus	0.74
Peripatopsis capensis	0.67
P. halfouri	0.39
P. leonine	0.31

[††] = in Trinidad, [**] = captive born,
[†] = in Jamaica, [*] = at embryonic stage

et al., 2000). Females tend to be heavier by 2.7 times and 50% longer than males (Monge-Najera, 1994). In *E. rowelli*, males are more abundant than females but only up to 200 mg size, beyond which females alone are present (Fig. 19.2A). In velvet worm species, in which female investment is high due ovoviviparity/viviparity, males do not need to be large to mate and are also not required after mating (Sunnucks et al., 2000). Despite a wide diversity in the modes of sperm transfer ranging from dermal insemination of spermatophores to specialized structures in males, most velvet worms exhibit internal fertilization. In spermatheca, the sperm remains alive and vigorously swims for 9.5 months (e.g. *E. rowelli*, Sunnucks et al., 2000). In oviparous species, development is commenced outside the female's body but within the protected shell. Conversely, development is completed and the hatched young ones are parturiated by the female. A female can harbor one developed and one under-developing batch of embryo in each uterus, indicating the independent use of their paired reproductive tract (e.g. *E. rowelli*). In *E. rowelli*, the reproductive tract holding an active sperm increases up to 80–100 mg male size but subsequently declines (Fig. 19.2A). Similarly, the sperm storing tract increases up to 300 mg size but decreases with subsequent increase in body size (Fig. 19.2B). Being a decisively important factor, fecundity increases with female body size (Fig. 19.2C).

According to Ruhberg (1990), the ovary consists of a pair of hollow tubes with thin, folded walls and follicular outgrowths projecting into the hemocoel. It is attached to the pericardial septum dorsal to the gut. Originating from the ovary, the oviduct lined with mucous cells, following receptacula seminis are dilated as uteri, which are joint to open into the vagina.

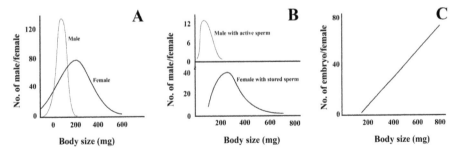

FIGURE 19.2

Euperipatoides rowelli: A. Number of males or females as a function of body size. B. Number of males with active sperm and females with stored sperm as a function of body size. C. Lifetime fecundity as a function of body size (modified and compiled from Sunnucks et al., 2000).

Oviparity and Viviparity

With direct development, onychophorans exhibit oviparity, ovoviviparity and viviparity. Their egg size ranges from 0.03 mm in viviparous yolkless eggs of neotropical peripatids to 0.5–0.15 mm in ovoviviparous yolky eggs (e.g. *Eoperipatus*, *Peripatoides*, *Paraperipatus*) and to 2.0 mm in oviparous yolky eggs of *Ooperipatus* and *Ooperipatellus* (see Eriksson and Tailt, 2012, Mayer et al., 2015). Notably, oviparity occurs in some peripatopsids only and viviparity in peripatids only (Table 19.3). In them, two patterns of cleavage are

TABLE 19.3

Oviparous, ovoviviparous and viviparous onychophorans (modified from Mayer et al., 2015)

Development Mode	Embryo-maternal Interaction	Habitat
Peripatopsidae		
Oviparity	Yolky eggs – vitelline membrane + thick sculptured chorion	Australia, New Zealand, e.g. *Euperipatoides komagigrensis*
Lecithotrophic matrotrophy **ovoviviparity**	Retention of yolky eggs in uteri-persistence of vitelline membrane + chorion – limited matrotrophy	Australia e.g. *Euperipatoides rowelli*
Matrotrophic **ovoviviparity**	Retention of yolkless eggs in uteri-persistence of vitelline membrane but no chorion. Maternally nourished with no placentation	Indonesia, South Africa, Chile e.g. *Peripatopsis*
Peripatidae		
Lecithotrophic **ovoviviparity**	Retention of yolky eggs in uteri. Vitelline membrane and chorion persist. No maternal nutrition	South Africa, Australia, New Zealand e.g. *Eoperipatus*
Placental matrotrophic **viviparity**	Small yolkless eggs exclusively nourished maternally through placenta: egg envelopes are absent	All neotropic peripatids

recognized: (i) discoidal meroblastic cleavage and (ii) holoblastic cleavage. In placental viviparous species with small yolkless eggs, cleavage is total and equal; from further divisions, it generates a morula, which becomes attached to the internal wall. Further cell divisions result in a coeloblastula. Specialized cells establish the embryonic placentation. Gastrulation occurs by ingression of endomesodermal cells into the blastocoels. In others, the described modes of gastrulation differ even between closely related species. These are elaborated by Mayer et al. (2015).

20

Pentastomida

Introduction

The 144 speciose (Veterian Key) pentastomids are elongated, flattened or cylindrical hemocoelomatic segmented tongue worms. Their body is divided into a cephalothorax and trunk. It is covered with a soft, porous, chitinous cuticle, which is molted occasionally. Their crossly striated muscles are arranged into longitudinal and circular layers. Internal organs are suspended in the hemolymph contained in the hemocoelom. The ventral mouth surrounded by a chitinous buccal ring leads into the pharyngeal pump, esophagus and a straight gut opening as an anus (Fig. 20.1C). Respiratory and excretory systems are absent. In all, they have eight pairs of ventral

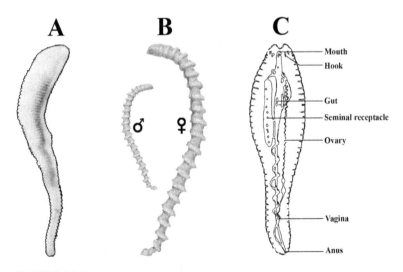

FIGURE 20.1

A. *Linguatula* (redrawn from Dr. M. Tavassoli, Faculty of Veterinary Medicine, Urmia University, Iran). B. Male and female of *Armllifer* sp (from Wikipedia). C. Anatomy of *Linguatula* (from Biocyclopedia).

ganglia, of which the first two pairs constitute the cerebral mass. Their size ranges from 80 mm to 4 cm (e.g. *Armillifer armillatus*, Ali et al., 1981).

Taxonomy and Distribution

Pentastomids include only one class: Eupentastomida comprising 3 orders, 7 families and 25 genera and 144 species (Veterian Key). They are mostly parasites in the respiratory system of reptiles; as a consequence, their geographic distribution is restricted to the tropics and subtropics (Christoffersen and de Assis, 2015). With direct development, they are also found in polar and sub-polar regions. *Linguatula serrata*, with the life cycle restricted to mammals, has become cosmopolitan.

Reproduction and Life Cycle

Pentastomids are gonochores (Fig. 20.1B) and reproduce sexually. Their male genital system consists of an unpaired testis, an unpaired (cephalobaenida) or paired (Raillietiellida) seminal vesicle(s), which is located between vas differentia and ejaculatory duct. The female reproductive system is complex. The ovary may be a paired organ, as it is indicated by a division anteriorly or posteriorly; spermathecae are situated between the oviducts and uteri, which terminate in a common vagina. Spermatheca store sperm and provide a site for continuous fertilization of oocytes. Banaja et al. (1976) found that after the 66th day following infection of the gull host with *Reighardia sternae*, the semelparous male copulates with the still juvenile female and dies. The females receive no further insemination during her lifetime. After 116 days, the fertilized female produces her lifetime fecundity of 2,000 eggs. No information is available on early embryonic development, which occurs within the female's uterus. Their embryos are reported to show metameric coelomic pouches and jointed appendages (Self, 1990). Embryonic development occurs within a series of egg membranes, which are secreted by a glandular organ. It is completed within the egg prior to hatching. The nymphal larva hatched from the egg consists of 4-segmented head and a 3-segmented trunk. Two more trunk segments are added during the post embryonic development following molting. The hatched larva is motile and capable of penetrating the intermediate/definitive host. It attains sexual maturity on reaching the respiratory system of the DH. The life cycle does not involve an intermediate host *Reighardia sternae* but it does involve one or more host (IHs). Figure 20.2 illustrates representative life cycles involving

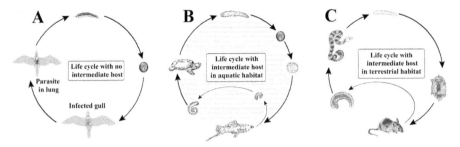

FIGURE 20.2

Representative life cycles of pentastomids (e.g. *Reighordia sternae*): A. Involving no intermediate host, B. involving an intermediate host in aquatic ecosystem and C. involving an intermediate host in terrestrial ecosystem (e.g. *Raillietiella scincoides*).

no host (Fig. 20.2A) and others involving an intermediate host in aquatic (Fig. 20.2B) and terrestrial (Fig. 20.2C) ecosystem.

Host Specificity and Prevalence

Pentastomids are generalists. For example, the treefrog *Litoria caerulea*, the geckos *Hemidactylus frenatus* and *Gehyra australis* as well as the toad *Rhinella marina* may serve as definitive host for *Raillietiella frenata* (Kelehear et al., 2014). *R. orientalis* can also be hosted by as many as six host species (Table 20.1). The prevalence of *R. orientalis* increases from 7.1 in *Dendrelaphis punctulatus* to 100% in *Demansia papuensis*; in them, intensity also increased from

TABLE 20.1

Prevalence and intensity of *Raillietiella orientalis* and in Australian snakes (condensed from Kelehear et al., 2014)

Snake Host Species	*Raillietiella orientalis*		*Waddycephalus* spp	
	Prevalence (%)	Intensity (no./host)	Prevalence (%)	Intensity (no./host)
Dendrelaphis punctulatus	7.1	1.0	78.6	3.9
Liasis fuscus	14.3	1.0	-	-
Acanthopsis praelongus	40.0	3.8	1.0	1.0
Tropidonophis mairii	41.7	2.8	1.0	1.0
Demansia papuensis	100.0	8.0	-	-
D. vestigata	100.0	12.8	21.1	1.3
Boiga irregularis	0	0	0	0
Antaresid childreni	0	0	0	0

1 parasite/host to 12.8 parasite/host in *Dem. vestigata*. In *Den. punctulatus*, *Waddycephalus* spp are more prevalent (79%) than *R. orientalis* (7.1%). The latter is more prevalent in *Dem. papuensis* than *Waddycephalus* spp. However the presence of *Dem. papuensis* in *R. orientalis* seems to exclude *Waddycephalus* spp. The reverse is true for *Den. punctulatus*. More information is required prior to generalizing that one species of pentastomid exclude the other.

Part E
Schizocoelomata: Lophophorata

This clad includes Entoprocta, Phoronida, Bryozoa and Brachiopoda. It is characterized by the filter-feeding device, the lophophore, U-shaped digestive tract, well developed coelom, radial cleavage, indirect life cycle and potency for regeneration and/or clonal multiplication. Entoprocts are characterized by a special determinate cleavage pattern. Being sessile, most of them are hermaphrodites, although some especially entoprocts are gonochores. Barring the speciose (5,700 species) Bryozoa, others are less speciose, each with a few hundred species.

21

Entoprocta

Introduction

The 200 speciose (Nielsen, 2013) endoprocts are minute (100 μm to < 5 mm, Hyman, 1951b), sessile, solitary or colonial, suspension feeding 'kamptozoans' with an anterior ciliated lophophore. Their body consists of a bowl-shaped calyx with a crown of 8–30 tentacles around a concave area, a long cylindrical muscular stalk, which may have bulbous enlargements along its length, and allows flexible body movements. The stalk is attached to the substratum directly by an adhesive disk in solitary species but in colonial species, it is joined with a ramifying cylindrical stolon with incomplete septa between sections with zooids and sections without zooids. Their body is covered by a one-layered epithelium with microvilli penetrating layers of crossing collagenous filaments. The epithelium is partly covered by a cuticle, which is reduced or lacking on the inner side of the tentacles and atrium, facilitating respiratory exchange. The body wall consists of a musculature comprising ring, diagonal and longitudinal muscles. The calyx contains a U-shaped ciliated gut with a flattened funnel-shaped esophagus, a large stomach, a conical intestine and a tubular rectum. The mouth is located at the base of the tentacles on the anterior side of the atrium and the anus on the opposite side (Fig. 21.1). Most colonial species have a pulsating 'star-cell complex' at the narrow transition between the calyx and stalk. Within the calyx reside the digestive tract, protonephridia, nerves and gonads. A dumbbell-shaped neural ganglion is located at the upper side of the stomach near the esophagus; peripheral nerves are extended to the other regions (Mariscal, 1975, Nielsen, 2013, Wanninger, 2015b).

Taxonomy and Distribution

In 1971, the phylum Entoprocta was indicated to have 60 described species by Hyman (1951b) and 120 species by Mariscal (1975). Hence, the number

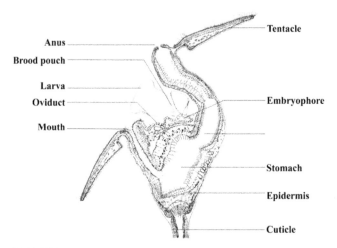

FIGURE 21.1

Free hand drawing of *Barentsia benedeni* female calyx showing the
anatomy (redrawn from Mariscal, 1975).

of new erected species increased at the rate of 2.5/y during 1951–1974 and
subsequently 2.1/y (Nielsen, 2013). It is likely that the number of described
species remains around 200 species (Wanninger, 2015b). Though the number
of species and families has changed, this account is based on the classification
by Mariscal (1975, see also Hyman, 1951b), as it provides information on
sexual reproduction and clonal multiplication. Accordingly, the phylum
includes three families in 13 genera. Of them, the members belonging to
the family Loxosomatidae in 5 genera are solitary (Fig. 21.2). Among others,
Pedicellinidae with 7 genera comprises colonial species (Table 21.1). Both
loxosomatids and pedicellinids are exclusively marine habitants. Two species
in the family Urnatellidae are also colonial but their distribution is restricted
to freshwater. Some contrasting features between the solitary and colonial
ectoprocts are listed in Table 21.2.

Reproductive Biology

Of ~ 103 entoproct species examined, information on sexuality is available
for 55 only; of which 45 are gonochorics and 10 hermaphrodites, i.e. 82%
of them are all gonochores (Table 21.2). In gonochores, females can be
distinguished from males, only when embryos are visible in the brood pouch
of the females or in species with differently colored gonads visible through
the transparent body wall. In *Loxosoma annelidicola* and *L. pectinaricola*,
females are smaller than males. But males are smaller than females in

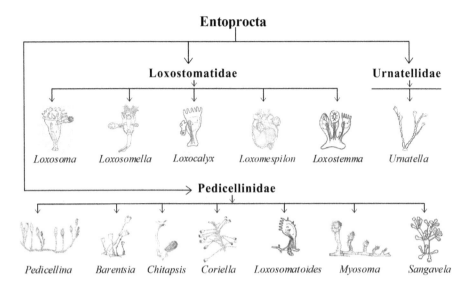

FIGURE 21.2

Classification of Ectoprocta (free hand drawings from Emschermann, 1971, Mariscal, 1975, Borisanova and Potanina, 2016, Wikipedia, *Chitapsis*, after Annandale, 1916, *Pedicellina*, after Ehlers, 1890).

TABLE 21.1

Contrasting features between the solitary loxostomatids and colonial pedicenellids (condensed from Ruppert et al., 2004, ITIS Standard Report, 2006)

Features	Solitary Loxostomatids	Colonial Pedicenellids
Septum between calyx and stalk	Present	Absent
Star-cell complex	Present	Absent
Anus on cone	Absent	Present
Stolons present	Absent	Present
Segmented stems	Absent	Present

L. cirriferum. Interestingly, both female and male individuals are produced along a single stolon (Mariscal, 1975). In both males and females, a pair of saccular gonad is located between the stomach and atrial cavity wall. The gonads open into the brood pouch by a single straight gonoduct (Fig. 21.3A, B). In hermaphrodites, the ovaries are located anterior to the testes and they shed the gametes through a common gonoduct. The Primordial Germ Cells (PGCs) are located adjacent to the respective ducts and give rise to oogonia and spermatogonia following repeated divisions. In *L. rhodnicola*, the mature filiform sperm measures ~ 70 μm in length and ~ 1 μm in diameter. The mature ovum is spherical and measures 80 μm in diameter. In *Loxosomella*

TABLE 21.2

Sexuality, number of embryos and clonals in known species of Entoprocta (condensed from Mariscal, 1975). UK = Unknown

Genus	Species (no)			Sexual Reproduction		Clonal Multiplication	
	Total	Gonochore	Hermaphrodite	Incidence	Number	Incidence	Number
Loxosomatidae							
Loxosoma	24	13	1	10	11	21	3.0
Loxosomella	28	14	1	22	5	25	5.0
Loxocalyx	12	3	2	3	3	11	2.5
Loxomespilon	1	1	0	1	6	1	4.0
Loxostemma	1	-	0	1	1	1	1.5
Subtotal/Average		31	4	37/7.4	26/5.2	59/11.8	16/3.2
Pedicellinidae							
Barentsia	12	5	-	3	6	1	UK
Pedicellina	10 + 4	5	5	UK	UK	UK	UK
Chitapsis	1	-	-	UK	UK	UK	UK
Coriella	3	-	-	UK	UK	UK	UK
Loxosomatoides	2	1	-	UK	UK	UK	UK
Myosoma	2	2	0	UK	UK	UK	UK
Sangavela	1	0	1	UK	UK	UK	UK
Urnatellidae							
Urnatellia	2	1	1	UK	UK	UK	UK
Total	103	45	10	40	-	60	-

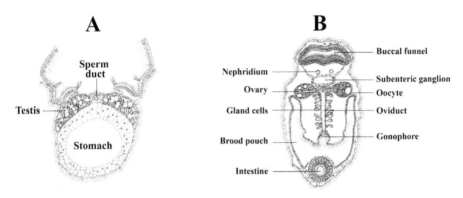

FIGURE 21.3

Transverse section of the calyx to show the reproductive system in (A) male and (B) female (free hand drawings from Mariscal, 1975).

ovipara, the zygote is very small 6 µm and increases in size at the expense of nutrients drawn from the surrounding follicular cells.

Cleavage is total and unequal, and is spiral determinate type (Fig. 21.4A–D). Following the third cleavage, four micromeres are recognizable at the animal pole and four macromeres at the vegetal pole. At the 56-cell stage, notably,

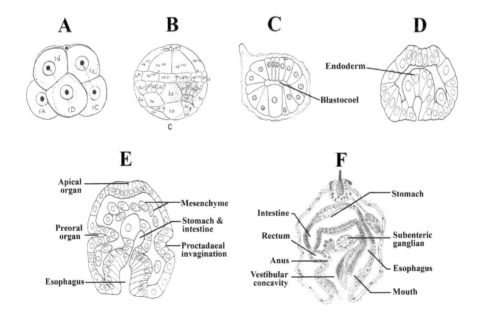

FIGURE 21.4

Development in *Pedicellina cernua*: A. 8-cell stage, B. 56-cell stage side view, C. the same as section through the blastula, D. 130-cell stage, E. Proctodaeum forming stage, F. almost complete stage (free hand drawings from Marcus, 1939).

the 4 D cell is the mesentoblast cell, which gives rise to the mesoderm cells that shall subsequently form the endomesoderm. Based on the yolk present in the egg, two types of embryonic and larval development are recognized. In type 1, the yolky eggs are retained in the atrium but with no transfer of maternal nutrients to embryos. In them, the third cleavage is dexitropic. With further cleavage, 16 equal sized blastomeres are produced. However, different sizes of them are produced following the fifth cleavage. Subsequently, a small blastocoel appears. At 100–120 cell-stages, gastrulation occurs by invagination. A proctodaeum is developed by posterior invagination and a frontal organ from the thickened ectoderm at the anterior. At this stage, the 4 D cell is divided to form a pair of lateral rows of mesodermal cells. Then the embryo is elongated and egg membrane disappears. With subsequent development, a fully grown larva starts feeding prior to its release from the atrium. Following its release, the planktotrophic trochophore passes through a long pelagic phase. In type 2, the area around the apical organ at the 16-cell stage comes in contact with the thickened epithelium of the brood pouch, which forms the placenta. Then the embryo increases considerably in size due to transfer of maternal nutrients. For example, the size of *Loxosomella vivipara* egg increases from 10 μm to 200 μm in an embryo. However, there are wide variations in development among the lecithotrophic *Loxosomella* larvae. Most of the larvae have a well developed gut but the gut is totally absent in *L. vivipara* larvae. Still, the type 2 larvae are lecithotrophic and pass through a short pelagic phase. For more details on metamorphosis and settlement, Nielsen (1990) and Wanninger (2015b) may be consulted.

Clonal Multiplication

Colonial entoprocts often degenerate and shed their old or injured calyces, which are regenerated by a process similar to budding in stolon. In solitary species, regeneration is reported from a few species only (e.g. *Loxosomella antertica*, see Iseto, 2017). However, clonal multiplication is ubiquitous among other entoprocts (Table 21.2, Hyman, 1951b). It is so prevalent that it has led to rarity of males and/or reduction in number of sexually reproducing individuals. For example, males are rare in *Loxocalyx neapolitanus, Loxosomella kefersteinii, L. crassicauda* and *Loxosoma rhodnicola* (Mariscal, 1975). From a 2-year long survey on sponge-inhabiting 2,300 individuals of *Loxosomella plakorticola*, Sugiyama et al. (2010) reported that sexual reproduction is limited to 16 (0.7%) individuals only.

In all entoprocts, budding is the most common mode of clonal multiplication. Essentially, it is an ectodermal (see Skold, 2009) or mesodermal (Mariscal, 1975) process. However, the formation of encapsulated resting hibernacula is another mode of clonal multiplication (e.g. *Barentsia benedeni*,

B. matsushimana, Konno, 1971). The hibernacula consist of large thickened finger-like outgrowths from the base of a stolon, which, with a change of environmental conditions, is capable of producing new individuals. Similarly, the freshwater *Urnatella* produce statoblasts that are capable of overwintering and subsequently, regenerating new individuals (Mariscal, 1975). In loxosomatids, buds arise on the sides of the calyx near its oral end in a pair of bilaterally symmetrical area (Fig. 21.5A). They may appear simultaneously or alternatively on the two sides. At a time, a single or a bunch consisting of 6–12 (e.g. *Loxosomella obesa, L. kefersteinii, Loxosoma davenporti*, see Mariscal, 1975) and up to 15 buds may arise, as in *Loxosomella variens* (Nielsen, 1964). In pedicellinids, buds are produced by stolons and stalks but not by calyx. The stolons are subdivided by cuticular septa; the fertile interseptal stalk-bearing bud(s) alternates with the sterile section devoid of a stalk (Fig. 21.5B). As the stolon creeps along the substratum, it gives rise to buds at right angles to its surface behind the advancing tip. Stalks can also bud at the muscularly thickened bead-like swellings (Fig. 21.5C). In *Urnatella*, buds appear from the beads; with proliferation of secondary buds, calyces are accumulated at the distal end of the colony (Fig. 21.5D). Precocious budding is also known in some entoprocts (e.g. *Loxosomella vivipara*). One to two adults are developed within a larva being brooded. In *L. vivipara*, up to four generations may be associated with a single parent: (i) mother, (ii) larva developing in parent,

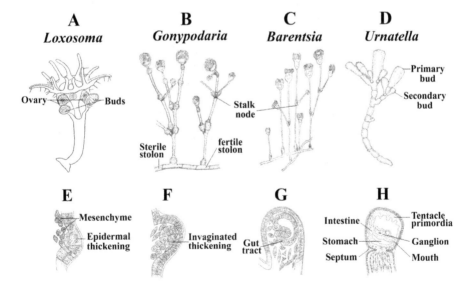

FIGURE 21.5

Buds in *Loxosoma* (after Atkins, 1932), primary and secondary buds in *Urnatella* (after Devanport, 1893), stalk nodes, fertile and sterile stolons on *Gonypodaria* (after Robertson, 1900) and *Barentsia* (from Hyman, 1951b). A–D. Transverse sections showing budding stages in *Pedicellina* calyx (after Marcus, 1939).

(iii) buds developing within the larva and (iv) buds developing on the bud contained within the larva (see Mariscal, 1975).

Essentially, the budding process is common to both solitary and colonial entoprocts. It begins with an increased rate of divisions in epithelium, pushing cells against each other. Thereby, it causes thickening and formation of ectodermal outpocketing of the epidermis, which is then constricted into inner and outer vesicles (Fig. 21.5E–H). The migrated mesodermal cells in the inner one invaginate to form the gut primordium. At the same time, the gonads, muscles and other mesodermal organs merge from the parental migrated undifferentiated mesenchymal cells. The tentacles, body wall and subenteric ganglion all are developed from the ectodermal precursors. Apparently, clonal multiplication is accomplished from the ectoderm and mesoderm alone (Hyman, 1951b, Mariscal, 1975). In *Arthropodaria*, removal of a fraction of the tentacular crown including the tentacular membrane readily regenerated the crown within a few days. However, the excision of the entire membernae (margin) of the calyx resulted in death. Any joint (node) of the stalk also regenerated the clayx. But the fertile fraction of the stolon alone revived the colony of *Arthropodaria* (Nasanov, 1926). Evidently, the multipotent stem cells for the calyx regeneration are retained in the tentacular membrane and at the stalk nodes. But the totipotent stem cells that can regenerate the entire colony are retained in the fertile fraction of the stolon alone.

Among colonial entoprocts, *Barentsia benedeni, B. graniculata, B. gracilis, B. macropus, Pedicellina americana, P. compacta, P. glabra, P. hildegradae, P. hispida* (see Iseto, 2017) and *Urnatella indica* are gonochores. *P. echinata, B. discreta* (Mukai and Makioka, 1980) and *B. aggregata* are hermaphrodites. Hence, it may be interesting to know whether the hermaphroditic species generate gonochoric clones and the gonochore hermaphorditic clones, i.e. whether the PGCs in the entoprocts are retained bisexual potency. Atkins (1932) reported the doubtful production of female buds by males and the reverse. In the absence of reliable publications, a single publication by Jagersten (1964) seems to indicate the heterogametic sex determination in *Pedicellina cernua*. In it, 50% of the buds in the parent were sexually matured with well developed testis filled with sperm. Presumably, the remaining 50% buds were females.

22

Phoronida

Introduction

According to Hyman (1959), Phoronida are slender, colorless or transparent, solitary or aggregated, tubiculous, deuterostomid worms bearing a crown of tentacles. They secrete a transparent sticky fluid, which forms the tube on contact with water. Their size ranges from 6 mm (length) with 18–25 tentacles (e.g. *Phoronis ovalis*) to 200 mm with up to 300 tentacles (e.g. *Phoronopsis viridis*). However, most of them measure > 100 mm (Fig. 22.1A, B). Their body consists of an epistome, anterior tentacular crown and a long slender

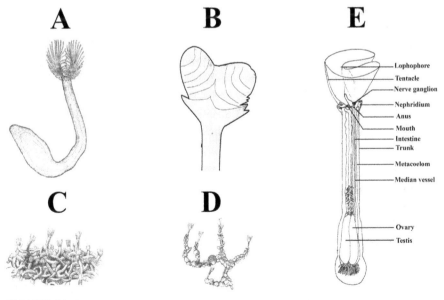

FIGURE 22.1

A. *Phoronis architecta* (after Wilson, 1881). B. *Phoronopsis californicus* (from http://when-why-how.com). C–D. Aggregations of *Phoronis psammophila* and *Phoronis hippocrepia* (after Wilson, 1881). E. Anatomy of a phoronid (free hand drawing from Emig, 1982).

trunk (Santagata, 2015a). The tentacles are borne in a single row along the double ridges of the body wall convexly curved into the shape of a crescent or horseshoe (Fig. 22.1C). The number of tentacles increases with age (e.g. *Ph. mulleri*). The slender tentacles are hollow and ciliated. Their body wall consists of an epidermis, basement membrane, thin-layers of circular and longitudinal muscles. Internally, it is lined by peritoneum, evident as a nucleated syncytial layer. In addition to the wall, the tentacles are covered by a cuticular layer. Their true coelomic cavity, lined by mesoderm, is divided into protocoel in the epistome, mesocoel in the lophophore and metacoel in the trunk (Santagata, 2015a). The metacoel contains colorless fluid, in which coelomocytes, phagocytic amoebocytes, basophilic granulocytes and non-motile eiosinophils are found. The U-shaped digestive tract extends throughout the trunk, and is recurved and terminates in an anus located outside the tentacles. A respiratory system is absent. But a closed circulatory system is present. Their colorless blood contains red blood corpuscles. A pair of metanephridium with their ducts is located at the anterior end of the metacoel. The 'brain' consists of a group of neuronal cells concentrated between the mouth and anus. Neural cells and fibers are sporadically distributed throughout the surface of the trunk epithelium. The most centralized neuronal structure is the giant nerve fiber embedded in epithelium of the anterior trunk (Hyman, 1959, Santagata, 2015a).

Taxonomy and Distribution

The phoronids are classified into two genera: *Phoronis* and *Phoronopsis*. In 1982, Emig indicated the presence of 10 species; at present, 23 species are known (Santagata, 2015a), i.e. newly erected species is increased at the rate of 0.4 species/y. The phoronids are exclusively marine inhabitants and are found sparingly over a wide geographical range, especially in European waters. They are limited to the upper littoral zone.

Reproductive Biology

The phoronids can be gonochores or hermaphrodites. In the former, sex cannot be distinguished externally except by the ripe gonad color visible through the body wall. The posteriorly located gonads are loose, indefinite masses, closely attached to the lateral vessel and capillary caeca (Fig. 22.2A). In hermaphrodite species, the ovary is located on the dorsal side of the vessel and the testis to its ventral side but the relation can be reversed, as in *Phoronis pallida*. The ripe gametes are shed into the coelom, where fertilization occurs

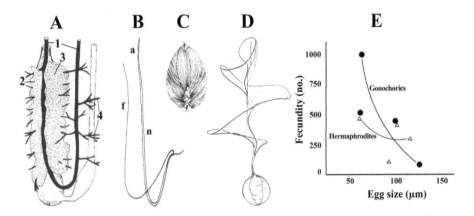

FIGURE 22.2

A. Posterior part of *Phoronis australis*, 1 = lateral vessel, 2 = testis, 3 = ovary, 4 = capillary caeca (after Benham, 1889). B. Spermatozoa, a = acrosome, n = nucleus, f = flagellum. Spermatophore C. A-type (e.g. *Phoronis ijimai*) and D. B-type (e.g. *Phoronopsis harmeri*) (after Zimmer, 1964). E. Egg size as function of fecundity in some phoronids (for details see Table 21.1).

in *Ph. mulleri*, *Ph. hippocrepia* and *Phoronobsis viridis*. In others, it occurs after they are shed directly into the sea water through nephridial openings. In continuously spawning species, the lophophoral concavity acts as a brood chamber, holding asynchronously developing eggs and embryos (Hyman, 1959). Mature spermatozoa (Fig. 22.2B) break away from the testis into the metacoelom and aggregate into a loose spherical mass near the nephridial funnel. In some species, the sperms are packed into an ovoid mass of spermatozoa (spermatophore type A, Fig. 22.2C). In others, the spermatophore type B (Fig. 22.2D) is packed, which is a fairly large spherical mass consisting of spermatozoa and a spiral float. Whereas, type A is collected by the nearest individuals, the latter can float away to longer distances (Emig, 1982).

Emig (1982) summarized sexuality, egg size, fecundity, spermatophore type and spawning periodicity for 10 phoronid species, for which information was then available. These values are reassembled in Table 22.1, which suggests that: (1) Of 10 species, 6 are gonochorics and the others are hermaphrodites. (2) All the four hermaphrodite species spawn continuously, whereas of the 2 gonochorics, one spawns periodically, while the other once in a lifetime. (3) In gonochores, the sperm may be released as such or packed in spermatophore type A, while hermaphrodites pack them in spermatophore type B. When fecundity values assembled by Emig were plotted against egg size (Fig. 22.2E), two inferences could be made of which the second is more important. (4a) As expected the fecundity decreases with increasing egg size, irrespective of sexuality and spawning periodicity. (4b) The trend for the relation between egg size and fecundity was at a relatively higher level for the gonochores. Arguably, with dual investment on reproductive system and

TABLE 22.1

Sexuality, egg size, fecundity, spermatophore types and spawning periodicity in some phoranids (modified from Emig, 1982). P = *Phoronopsis*, Ph = *Phoronis*, Sp = spermatophore

Species	Sexuality	Egg Size (μm)	Fecundity (no.)	Spawning	Sperm/ Spermatophore Type
Ph. pallida	♀	60	500	Continuous	-
Ph. hippocrepia	♀	93	100	Continuous	-
Ph. ijimai	♀	100	400	Continuous	-
Ph. australis	♀	115	300	Continuous	-
Ph. mulleri	♂, ♀	58	500	-	-
P. harmeri	♂, ♀	63	1000	-	Sp – B type
Ph. psammophila	♂, ♀	100	400	Periodic	-
Ph. ovalis	♂, ♀	125	40	One time	Spermatozoa
P. albomaculata	♂, ♀	100	-	-	Sp – B type
P. californicus	♂, ♀	-	-	-	Sp – B type

its maintenance, the hermaphrodites produce only half the number of eggs, as the gonochores can (cf *Ophyrotrocha*, Pandian, 2020).

Cleavage and Life Cycle

In most phoronids, fertilization is internal. Cleavage is total, almost equal and is of the biradial type (Fig. 22.3A). Following gastrulation, the cup-shaped archenteron extends into a tubular canal (Fig. 22.3B). Rapid growth of the ectoderm in the anterior region leads to the formation of a preoral lobe. The ectodermal cells at the posterior region join the intestine to establish the anus. From the anterior regions of the archenteron, cells are budded off as isolated mesoderm cells within the blastocoel. These cells give rise to the initial mesoderm for the development of the pre-oral lobe, where the protocoel appears. Difficulty encountered with rapidly developing mesoderm in the trunk has led to suggest that the mesoderm development in phoronids is a modified enterocoelous type. In the phoronids, the development is grouped into three types (Emig, 1982, 1990, Santagata, 2015a). In Type 1, in a less yolky small egg of ~ 60 μm (e.g. *Phoronis pallida*) embryonic development is rapid and limited to a duration of 5 days followed by a prolonged planktotrophic actinotrocha larva (Fig. 22.3C). In Type 2, the yolky (100 μm, e.g. *Phoronopsis albomaculata*) eggs and embryos are brooded in the parental tube up to metasomal stage and is followed by equally long pelagic/creeping larval stage and subsequent settlement on hard substratum by burrowing or

A. Early cleavage in type 2

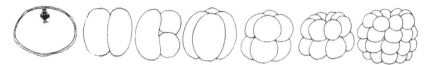

B. Gastrulation and organogenesis

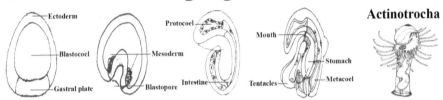

Actinotrocha

Ectoderm · Blastocoel · Gastral plate · Protocoel · Mesoderm · Blastopore · Intestine · Mouth · Stomach · Tentacles · Metacoel

C. Development types

Time 0 ——————— 5 ——————— 10 ——————— 15 ——————— 20 d

Type 1 ☆☆☆☆B☆☆☆☆G ——————————— *********M*

Type 2 ☆ B ☆ G☆☆☆☆☆☆☆☆☆☆☆☆☆——2.T——— MS —— BM ——— DV ——*********M*

Type 3 -☆ B ☆ G——————2.T——— MS —— BM ——— DV ——*******M*

FIGURE 22.3

Phoronidae: A. Early cleavages in Type 2 egg. B. Gastrulation and organogenesis. C. Development types in relation to time at the following stages: B = blastula, G = gastrula, MS = metasomal sac, BM = brood mass, DV = dorsal blood vessel and M = metamorphosis (free hand drawings from Emig, 1990).

encrusting. In Type 3, the yolky large egg (125 µm, e.g. *Phoronis ovalis*) is brooded in the nidanental glands for 7–8 days followed by 9–14 days non-feeding lecithotrophic larval stage and settlement in soft substrate. With indirect development, their life cycle includes embryonic development followed by planktotrophic/lecithotrophic larval stage.

Regeneration and Clonal Multiplication

Several authors have observed the casting off of the tentacular crown in natural populations of many phoronid species (Hyman, 1959). The act of casting off the crown is accomplished by a strong constriction of circular

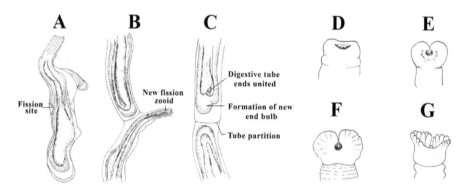

FIGURE 22.4

A–C. Stages of fission in *Phoronis ovalis* (after Harmer, 1917). D–G. Four stages of regeneration of tentacular crown. Note the formation of lophophore around the mouth (after Selys-Longchamps, 1907).

muscles (Fig. 22.4A), which heal the ends. In an experimental study, Marsden (1957) amputated *Phoronis vancouverensis* at levels of (i) tentacular crown alone, (ii) middle of the muscular tentacle, (iii) tentacular base and (iv) middle level through the trunk inclusive of reproductive vasoperitoneal tissue. In the first three, the wound was healed; the trunk with a part of a tentacle, the posterior ramet regenerates the anterior fraction but the anterior ramet fails to regenerate the posterior. Evidently, the potency for wound healing is retained within the circular muscles of both the tentacles and trunk. However, the totipotency is retained only in the reproductive vasoperitoneal tissue to regenerate a phoronid. Details on the regeneration process are reported by Schultz (1903). Accordingly, crescentric ridges appear over the esophagus in the decapitated trunk, increase in height, become subdivided into tentacles and thereby restore the lobophore (Fig. 22.4D–G). The two main longitudinal vessels fuse at once and promptly re-establish the circulation. The cut end of the intestine closes and reforms the anus by a simple fusion without a proctodeal invagination. The septum, coelomic cavities and nephridia all appear from peritoneal cells (Fig. 22.4A–C). Evidently, it is the mesoderm cells emerging from the archenteron that have retained the totipotent reproductive vasoperitoneal cells that have descended from the archenteron-based mesoderm cells, in type 2 regenerate the missing trunk components as well as the tentacles including the crown. This inference is also confirmed by the report by Selys-Longchamps (1907), who found that all the six pieces of the muscular (reproductive vasoperitoneal tissue) trunk regenerated into complete six new individuals.

23

Bryozoa

Introduction

The polyphyletic (Wood, 2010) Bryozoa inclusive of Ectoprocta are microscopic (a zooid measures 0.5 mm but 1–2 mm or even 4 mm in *Cylindroecium*), sessile, colonial, lophophorate coelomates. The body of an individual called zoarium consists of the tentacular lophophore protrusible through an orifice and trunk that are covered by an exoskeletal case termed zoecium or gelatinous material of its own secretion. The delicate ctenostome colony *Zoobotryon pellucidum* may reach a height of 45 cm and the gelatinous *Alcyonidium* up to 60–90 cm in length; *Membranipora membranacea*, found on fronds of the brown kelp *Laminaria*, extends up to 150 cm and contains 2.3 million zooids. The lophophore, identical to that of phoronids, consists of body wall extension subdivided distally into a single row of ciliated tentacles. With the presence of coelomic cavity, all parts of the lophophore are hollow. In marine bryozoans, the number of tentacles varies from 8 to 13 or even 40; the diameter of the spread-open lophophore is highly correlated with the tentacle length, and the tentacle number is in turn highly correlated with the diameter. These parameters of the lophophores, as feeding structures, have important consequences on growth and reproduction of bryozoans (Gordon et al., 1987). The tubiculous *Hypophorella expansa* is shown to have a special gnawing apparatus (Prots et al., 2019). The trunk coelom, lined by peritoneum, is separated from that of lophophore by an imperfect septum. The body wall is composed of living tissues collectively called the endocyst together with a non-living outer ectocyst; the endocyst is better known as the cystid (Hyman, 1959). In other words, each zooid in the colony has two basic components: an organ system or partially protrusible polypide and body wall or cystid, which encloses the entire polypide (Wood, 2010). In the simplest gymnolaemates, and ctenostomes, the wall consists of chitinous zoecium of various thicknesses, the underlying thin epidermis and the thin syncytial peritoneal network but not a continuous sheet of lining. The furnicular cords clothed with peritoneum attach the stomach caecum to the posterior zooidal wall. Calcification of the zoecium is the prevalent means

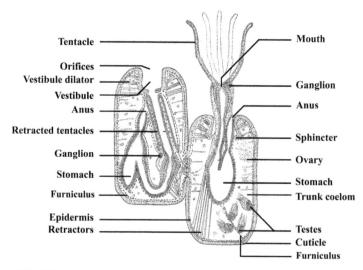

FIGURE 23.1

Schematic figure to show retracted and extended bryozoan zooids (free hand drawing from Marcus, 1926).

of fortifying the colony and is common among cheilostomes and universal in cyclostomes (Hyman, 1959). In the calcified zoecium, calcium carbonate is continuously deposited by the epidermis. The recurved digestive system commences with a simple round mouth opening at the base of the lophophore and terminates in an anus located at the root of the lophophore but on opposite side of the mouth. Various muscles associated with protrusion or retraction of the lophophore is shown in Fig. 23.1. The nervous system consists of plexus throughout the body wall and a main ganglion mass is encircled by the pharynx. Respiratory, circulatory and excretory systems are wanting. Bryozoans can also be economically important, (i) as many freshwater bryozoans serve as hosts of malacosporean Myxozoa that causes diseases to economically important fishes (Hartikinen and Okamura, 2019), (ii) as their statoblast dispersal is mediated through aquatic birds to over 757 km (Okamura et al., 2019) and (iii) as many bryozoans are considered as a promising source of anticancer drugs (Figuerola and Avila, 2019).

Taxonomy and Distribution

Taxonomy: The phylum Bryozoa comprises 5,700 species (Hondt, 2005). However, the current estimates run up to 8,000 living species and > 15,000 fossil species (Santagata, 2015b). The phylum was divided into two major classes by Hyman (1959). With elevation of suborder Stolonifera to a Class status, the phylum is presently divided into three major Classes namely (1) Phylactolaemata, (2) Gymnolaemata and (3) Stenolaemata. The first comprises a single order: Plumatellidae, the second two orders: Cheilostomata and Ctenostomata and the third a single order: Cyclostomata. Following Massard and Geimer (2008),[†] Bock and Gordon (2013), their general characters are summarized below:

Class: Phylactolaemata   Order: Plumatellida

Cylindrical zooids with a crescent lophophore and an epistome – Non-calcareous, muscular body wall facilitating eversion of lophophore – Coelom continuous between monomorphic zooids – New zooids arise by polypide replication – Production of special dormant statoblasts – Exclusively freshwater – 94 species in 24 genera and 10 families[†]

Plumatella repens
(Wikipedia) ~

Class: Gymnolaemata

Cylindrical or flattened zooids – Protrusible lophophore – Mostly **marine** bryozoans

1. Order: Cheilostomata

Calcified forms with opercula – Eggs brooded in ovicells – Jurassic to recent – Includes majority of living forms – 4,921 species in 129 families

Bugula foliolata

2. Order: Ctenostomata

Cylindrical to flat zooids – Terminal orifice closed by a pleated collar – No ooecia or avicularia – Jurassic to present – ~ 319 species in 32 families

Bowerbankia sp
(Tech University, Ostrova)

Class: Stenolaemata

Tubular zooids – Calcareous body wall without muscle fiber not used for eversion of lophophore – Polymorphic zooids separated by septa – New zooids produced by division of septa – Marine bryozoans – 543 species in 250 genera and 23 families

1. Order: Cyclostomata

Circular orifice and lophophore – Zooids interconnected
by open pores – Polyembryonic sexual reproduction in
special zooids – Ordovician to present

Entaloporoecia robusta

Geographic distribution: Gymnolaemates are known from all seas at all depths up to 6,000 m. *Criscia eburnea* is a circumarctic and circumboreal cyclostome (Hyman, 1959). More information is available on distribution of bryozoans in freshwater habitats. Understandably, the distribution of gelatinous phylactolaemates is somewhat limited to still waters (e.g. *Plumatella nitiens*), but plumatellids with a resistant cuticle also occur in swift streams (e.g. *P. emarginata*). The phylactolaemates thrive in well oxygeneated waters (6.3–7.4 ml O_2/l) and grow mostly in alkaline waters (pH 8.0–8.2); however, *Fredericella indica* (pH 4.9), *Plumatella fruiticosa* (pH 5.7) and *P. repens* (pH 6.3) are recorded from acidic waters (Wood, 2010) *Lophopodella capensis* thrive in alkaline waters of 8.3–9.1 pH in Central Africa. They are found in waters at 4–9°C up to 69°N. Their vertical distribution ranges from the altitudes of 3,840–4,150 m in Lake Titacaca, Bolivia (e.g. *Fredricella sultana*, *Stolella agilis*) down to 1,750 m depth in a Brazilian lake (Hyman, 1959). Some like *Electrapilosa* and *Conopeum reticulatum* are cosmopolitan, with distribution from tropical Indian to boreal waters (Menon, 1972). However, others have continentally restricted distribution. The abundant European *Plumatella fungosa* is rarely found in North America; *Fredericella indica* and *Plumatella indica*, commonly found in North America, are rare in Europe (Wood, 2010). Available information on species richness and geographic distribution of freshwater bryozoans is summarized in Table 23.1, from which the following may be inferred: (1) Of 94 species in 24 genera and 10 families, the phylactolaemates and gymnolaemates constitute 74 and 20 species, respectively. (2) In them, species number is 4.6/phylactolaemate genus and 3.9/gymnolaemate genus. (3) About 50% species are confined to one geographical region alone. (4) The Arctic and tropics are the richest geographic regions.

Colonial Forms and Polymorphism

Clonal multiplication and the consequent colony formation have reached a climax among Bryozoa. The simple *Aetea* colony consists of a succession of single zooids partly erect and partly adherent to the substratum (Fig. 23.2A). In other simple colonies, zooids are borne on erect or creeping stolon in pairs (e.g. *Aeverrillia*, Fig. 23.2B) or on clusters (e.g. *Bowerbankia*, Fig. 23.2C) or on irregular rows (e.g. *Visicularia*, Fig. 23.2D). A stolon is a tubular extension of the body wall and is divided by perforated partitions called septa or nodes into lengths termed as internodes that are regarded as highly altered zooids.

TABLE 23.1

Distribution of number of species, genera (in square brackets) and number of species confined to one geographical region only (in brackets). PA = Palaearctic, NA = Nearctic, NT = Neotropical, AT = Africotropical, OL = Oriental, AU = Australasian, PAC = Pacific Oceanic Islands, World = total of species per taxon (compiled from Massard and Geimer, 2008)

Taxon	PA	NA	NT	AT	OL	AU	PAC	World
Phylactolaemata								
Fredericellidae	4 [2]	4 [1]	3 (1) [1]	2 [1]	4 [2]	2 [1]	-	6 (1) [2]
Plumatellidae	24 (5) [6]	20 (4) [4]	19 (6) [3]	14 (3) [6]	30 (13) [9]	11 (2) [2]	2 [1]	59 (33) [9]
Pectinatellidae	1 [1]	1 [1]	1 [1]	-	1 [1]	-	-	1 (0) [1]
Cristatellidae	1 [1]	1 [1]	-	-	-	-	-	1 (0) [1]
Lophopodidae	4 [3]	2 [2]	1 [1]	4 (1) [1]	4 (1) [2]	1 [1]	-	7 (2) [3]
Subtotal	34 (5) [13]	28 (4) [9]	24 (7) [6]	20 (4) [8]	39 (14) [14]	14 (2) [4]	2 (0) [1]	74 (36) [16]
Gymnolaemata								
Victorellidae	5 (4) [3]	2 (1) [2]	1 [1]	1 (1) [1]	2 (1) [1]	1 [1]	-	8 (7) [4]
Pottsiellidae	-	1 [1]	1 [1]	-	-	-	-	1 (0) [1]
Paludicellidae	1 [1]	1 [1]	2 [1]	-	1 [1]	1 [1]	-	2 (0) [1]
Arachnidiidae	1 [1]	-	-	1 [1]	-	-	-	1 (0) [1]
Hislopiidae	3 (2) [1]	-	2 (1) [1]	-	5 (3) [1]	-	-	8 (6) [1]
Subtotal	10 (6) [6]	4 (1) [4]	6 (1) [4]	2 (1) [2]	8 (4) [3]	2 [2]	0 (0)	20 (13) [8]
Total	44 (11) [19]	32 (5) [13]	30 (8) [10]	22 (5) [10]	47 (18) [17]	16 (2) [6]	2 (0) [1]	94 (49) [24]

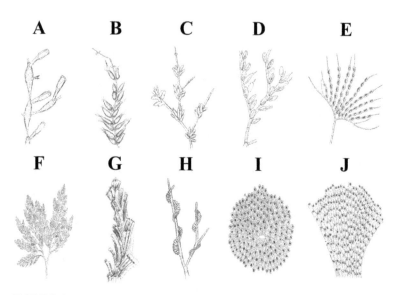

FIGURE 23.2

Forms of bryozoans colonies. A. *Aetea* (after Busk, 1886), B. *Aeverrillia* (after Rogick and Croasdale), C. *Bowerbankia* (after Marcus, 1938), D. *Vesicularia* (after Busk, 1886), E. *Cothurnicella* (after Levinsen, 1909), F. *Bugula*, G. Tubuliporid colony (after Busk, 1886), H. *Amathia*, I. *Cryptosula*, J. *Schizoporella unicornis* (redrawn from Hyman, 1959).

In *Cothurnicella*, each internode gives rise to a single file of zooids (Fig. 23.2E). The non-stoloniferous colonies of gymnolaemates consist of a continuous succession of zooids fused to each other. In non-calcified colonies like *Bugula*, the colony may form plant-like branches (Fig. 23.2F). Each branch may consist of a succession of zooids in uni-, bi- or multi-serial arrangement with zooids facing a single direction or any direction (Fig. 23.2G, H). With increasing calcification, the gymnolaemate colonies become compact and shorter. In reticulate colonies, the branches anastomose and are lace-like in appearance. The majority of calcified gymnolaemates are incrusted and occur as thin sheet on shells and other objects. The incrusting colony may radiate from a central point (e.g. *Cryptosula*, Fig. 23.2I) or may spread like a fan (e.g. *Schizoporella unicornis*, Fig. 23.2J).

In a colony, zooids can be polymorphic and reach their most varied expression in cheilostomes. Typical zooids with fully developed feeding polypides are called autozooids. Other types of zooids are termed collectively as heterozooids; the gonozooids are characterized by the loss of nutritive function. The others are avicularia, vibracula, kenozooid and dwarf zooid. Silen (1977) designated male and female gonozooids as androzooids and gynozooids, respectively. Sexual dimorphism may introduce another dimension to polymorphism. Following Hageman (1983), Ostrovsky (2013) classified the dimorphism into four types. In type 1, gonozooids are morphologically similar in appearance to sterile autozooids. But in type 2, they differ from sterile autozooids through enlargement (e.g. *Reciprocus regalis*) or reduction (e.g. *Selenariopsis galerieli*), sometimes accompanied by changes in shape (e.g. *Quadriscutella papillata*). In type 3, gonozooids differ in polypide morphology. For example, gynozooids of *Thalamaporella californica* have only 14 tentacles, while androzooids and autozooids have 17. In type 4, gonochoric zooids are characterized by sexual dimorphism (e.g. *Celleporella hyalina*) expressed in modification of cystid, polypide or both. In *Selenaria maculata*, autozooids, androzooids and gynozooids differ from each other in size and shape of the cystid, and their location in the colony, as well.

Reproductive Biology

Sexuality

Bryozoans are colonial hermaphrodites with testes and ovaries developing either within the same zooid, i.e. zooidal hermaphroditism (Fig. 23.1) or in different zooids within the same colony, i.e. zooidal gonochorism. The following may serve as examples for various expression of sexuality: (1) The cyclostomatid *Lichenopora* consists of simultaneously functioning male, female and hermaphrodite zooids. (2) Some cheilostomes consists of sterile, male, female and hermaphrodite zooids, the latter being more common, as in *Tendra zostericola* or rare, as in *Chartella membranacea truncata* and *Carbasea indivisa*. (3) The cheilostome *Thalamoporella evelinae* contains sterile, male and female zooids with males five to six times as numerous as females, which can be distinguished with 14 tentacles from males with 17 much larger tentacles. (4) *Synnotum aegyptiacum* consists of successive pairs of zooids, one male and one female in each pair. (5) In the protandric *Alcyonidium duplex*, the female zooid develops an ovary from the zoecia previously occupied by testes. This indicates clearly the retention of bisexual potency by Primordial Germ Cells (PGCs). (6) In the cheilostome *Celleporella hyalina* (Fig. 23.3C), the male ratio increases, especially after the age of 120 days. (7) Some species of the genus *Crisia* are gonochoric, occurring as male and female colonies. (8) In laboratory culture, colonies of *Filicrisia geniculata* function either as males or females (Hyman, 1959, Ostrovsky, 2013).

Thanks to Ostrovsky (2013), information is available on sexuality of 149 bryozoan species. In them, the distribution of sex is grouped in four types, as indicated below:

1a.	Sterile + ♀	10	4a.	Sterile + ♀ + ⚥	3
1b.	Sterile + ♀ + ♂	25	4b.	Sterile + ♀ phase + ⚥	3
2.	Sterile + ⚥	32	4c.	Sterile + ♀ phase + ♂ + ⚥	2
3a.	Sterile + ♂ + ⚥	15	4d.	Sterile + ♀ phase + ♀ + ⚥	1
3b.	Sterile + ♂ phase + ⚥	5	4e.	Sterile + ♀ phase	53

From the list, the following may be inferred: (1) In the 149 species, sterile autozooids are always present. (2) The distribution of sex in the four major types clearly indicates the retention of bisexual potency by PGCs in bryozoans. (3) In them, gonochorism is limited to a combination of sterile + female gonozooids in 10 species and + 25 species with sterile + male + female. A combination of sterile + male gonozooids is not thus far reported. (4) All others are hermaphrodites, either with sterile autozooids alone in 32 species or in other combinations. Female phase (53 species + 2 + 1 = 56 species) or male phase (5 species) gonozooids occur in combination with sterile autozooids. In them too, the number of species with female phase gonozooids is more than those with male phase gonozooids. As a rule, congeneric species have the same sexuality. Nevertheless, intrageneric variants have also been noted: 2 and 4 in *Callopora*, 1 and 2 in *Bugula* and *Celleporella*, 1 and 4 in *Steninoporella*, 1 and 3 in *Schizomarvello*. (5) In a colony, hermaphroditism facilitates to manipulate resource allocation to either male or female gametes, depending on circumstances and thereby increases the chances of fertilization. The sequence, in which the gonozooids appear, varies; for most of the time, *Electra posidoniae* colony consists of sterile + hermaphrodite zooids and the males appear towards the end of reproduction. In fact, a bryozoan colony is a dynamic system, in which the gonads form, mature and function at different times at various zooidal generations. Interestingly, zooids may change sex or acquire it. In the overwintered colonies of *Chartella papyracea*, many of the former gynozooids lose the ovary in autumn and become androzooids in spring. After functioning as autozooids for 1–2 months, some of them switched to function as normal androzooids in *Celloporella hyalina*, *Antarctothoa bougainvillei* and *A. tongima*. A single reparative androzooid was formed inside the former autozooid cystid in *Antarctothoa* sp (Ostrovsky, 2013). It is notable that switching sex in all these instances are in a single direction, i.e. from autozooid/gynozooid to androzooid. Whereas, the autozooids and androzooids have lost the bisexual potency of the PGCs and are able to differentiate only into spermatogonia alone, the gynozooids have retained bisexual potency.

Although the gametes are formed in numerous individualistic zooids, bryozoan reproduces as an integral system. In *Selenaria maculata*, for example, male and female zooids are always formed at or near the periphery of the colony. The male zooids are located around the edge of the periphery, while female zooids are placed subperipherally. This structural arrangement prevents them from receiving sperms from the same colony. With frontal budding in cheilostomes, the brooding zooids with embryos are positioned centrally facilitating fragmentation of the colony at the peripheral area. The evolution of polymorphism and sexual dimorphism is a striking feature for colonial integration. Notably, the progressive specialization of morphism of their cystid and/or polypide modules indicates that some of them function as an 'organ' of an individual in a colony. The formation of placental viviparity receiving nutrients from the autozooids is another striking example for colonial integration. The extensive network of furnicular cords are suggested to facilitate hormonal regulation to ensure different levels of colonial integration and synchronization of reproductive activity (Ostrovksy, 2013).

Apart from the combination of sex types, the quantitative aspects of autozooids and gonozooids as well as sex ratio may be modulated by an internal factor like the colony age and external factors like food availability. The cheilostome *Celleporella hyalina* exhibits complete partitioning between somatic (autozooids) and sexual (gonozooids) functions. Their dimorphic gonozooids are frontally budded from an underlying layer of autozooids. Hughes (1989) manipulated water flow rate and thereby food supply level, and estimated the colonial growth. On the 180th day, the number of zooids increased to 3,351, 2,152 and 1,604 zooids at the unrestricted, semi-restricted and restricted food supply, respectively. Of these zooids, approximately 0.63 and 0.37% zooids were feeding autozooids and gonozooids, irrespective of the changes in the food supply level (Table 23.2). Interestingly, it indicates the need for two autozooids to support a gonozooid. Of the 0.37% zooids allocated to gonozooids, the male and female gonozooids received equal (50%) share, irrespective of the changes in the food supply level. Clearly, changes in the food supply level may not alter the allocation for sexual function or the ratio between male and female function at least at the initial stage. Clonal sex ratio proved largely insensitive to differences in food supply and season. However, an earlier publication by Hughes and Hughes (1986) indicated that the male ratio, which were lower than 0.5 initially (Fig. 23.3B), began to increase from 120th, 135th and 170th days at restricted, semi-restricted and unrestricted food supply. In protogynic bryozoans, the male ratio may increase with advancing age, as in *Celleporella hyalina* and the reverse may be true for protandrics. That gonozooids are generated at the cost of autozooid has been suggested by hypothetical consideration (Fig. 23.3A) as well as field observations (Fig. 23.3C). Hughes (1989) hypothesized that with advancing colonial age and consequent increase the number of autozooids, the clones 1, 2 and 3, growing to long (S_1), longer (S_2) and longest (S_3) durations may increase the proportion of gonozooids at the cost of autozooids (Fig. 23.3A). In the White Sea, Nekliudova et al. (2019) found contrasting trends for the proportion of autozooids and gonozooids in the overwintered *Cribrilina annulata* colony. This observation clearly shows that gonozooids are generated at the cost of feeding autozooids.

TABLE 23.2

Celloporella hyalina: Approximate numbers of zooids, autozooids, male and female gonozooids generated during 180 days under different food supply conditions (calculated from Hughes, 1989)

Food Supply	Total Zooids (no.)	Autozooids (%)*	Gonozooids (%)*	♂ Zooids (%)*	♀ Zooids (%)*
Unrestricted	3,351	0.63	0.37	0.180	0.190
Semi-restricted	2,152	0.63	0.37	0.185	0.185
Restricted	1,604	0.63	0.37	0.180	0.190

* = as % of total zooids

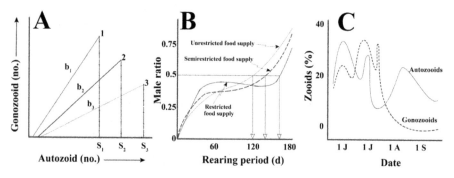

FIGURE 23.3

A. For the hypothetical clone (1, 2, 3), the number of gonozooids are regressed against that of autozooids. A negative correlation between the values of b and S indicates that development of gonozooids occurs at the cost of autozooids. B. Male ratio as a function of rearing period in *Celloporella hyalina* (compiled and modified from Hughes, 1989, Hughes and Hughes, 1986). C. Contrasting trends in proportion of autozooids and gonozooids of *Cribrilina annulata* during different calendar months in the White Sea (modified from Nekliudova et al., 2019).

Gonads and Gametes

In bryozoans, the male gonad lacks cell walls, ducts and accessory glands. Because of their 'loose' nature, it is called more appropriately as spermatogenic tissues. From the Primordial Germ Cells (PGCs), spermatogonia are formed after repeated mitotic divisions of PGCs in the cystid mesothelium that lines the body cavity and/or within the furniculus strands. Gymnolaematids have a single testis located on the basal wall or on the furniculus. But a pair of testes is reported from ctenostomes (e.g. *Alcyonidium mytili*) and cheilostomes (e.g. *Chartella membranaceo truncata*). The voluminous female gonad is more compact than that of the male. The number of ovaries is one in most bryozoans but sporadic incidences of two (e.g. *Bugula simplex*) and several (e.g. *Einhornia crustulenta*) ovaries are reported. The gonads may be located at different parts of the body of the zooid. The oocytes develop in pairs, one cell of the pair acting as a nurse cell or nahrzellen, which fuses with the proper oocyte, as soon as both reached a diameter of 20–30 μm, after which fertilization occurs. The mature ova or sperm first enter the body cavity and then are either released into the environment via gonopores or retained in the brood chambers (see Ostrovsky, 2013).

Based on (i) the type, size, number of oocytes and sequence of their maturation, (ii) the site and time of fertilization, (iii) the presence or absence of extra-embryonic nutrition during incubation and (iv) larval type, Ostrovsky (2013) recognized five reproductive patterns, of which the following three are described. In Pattern 1, oocytes synchronically mature in the ovary and the small oligolecithal eggs are fertilized in the ovarian cavity during or immediately after ovulation; the cohorts of these eggs are spawned into water. The embryos develop into planktotrophic cyphonautes larvae (Fig. 23.4A). Majority of the gymnolaemate cheilostomes (Suborder: Malagostegina) and some ctenostomes are oviparous and broadcast spawners. Their egg size ranges from 50 μm in *Membranipora serrilamella* to 100 μm in *M. membranacea*. Some of the variations in the egg size can be traced to imbibition of water, change to a round shape and/or measurement in live or fixed eggs. In these eggs, the oocyte in the ovary, receive reserved nutrients just to sustain embryonic development. In cheilostome and ctenostome broodcasters, oocyte production ranges from 10 to 60 per gynozooid at a given time. The duration of planktotrophic cyphonautes larvae ranges from 5 days in *Hislopia malayensis* to 4 weeks in *M. membranacea* and up to 6 weeks in *Alcyonidium mytili*. Figure 23.4A represents the typical life cycle in ctenostomes like *Alcyonidium nodosum* and *Farellarepens* and cheilostomes like *Electra* or *Membranipora*. Notably, the transition from reproductive Pattern 1 to other patterns is accompanied by asynchronous or sequential egg maturation, providing an adequate time interval to enlarge the egg size with more and more yolk deposition (Ostrovsky, 2013).

Pattern 2 is found in some Gymnolaemata. It is characterized by asynchronous maturation of the mesolecithal egg in the ovary, intra-ovarian fertilization and sequentially brooded embryos. The brooding may be external or internal. The sites of external brooding can be in the gymnocyst (e.g. *Victorella pavida*), the swollen base of tubular orificial collar (e.g. *Penetrantia*), simply sucked into the body wall (e.g. *Bulbella abscondita*), fastened to the outer surface of the diaphragm (e.g. *Alcyonidium duplex*), on or into the vestibule (e.g. *Cryptosula pallasiana*), into the tentacular sheath (e.g. *Flustrellidra hispida*), distal part of the coelom (e.g. *Nolella dilatata*) or in coelomic sac (e.g. *Baeniama gellanica*). In most cheilostomes, brooding occurs in a special external chamber known as ovicell or oecium that "perches like helmet or hood on the anterior end of some or all autozooids of a breeding colony" (Hyman, 1959). The ovicell is formed of two body-wall evaginations. The fertilized egg(s) are transferred into the ovicell; within it, the egg develops into a ciliated larva and escapes into the water. For an

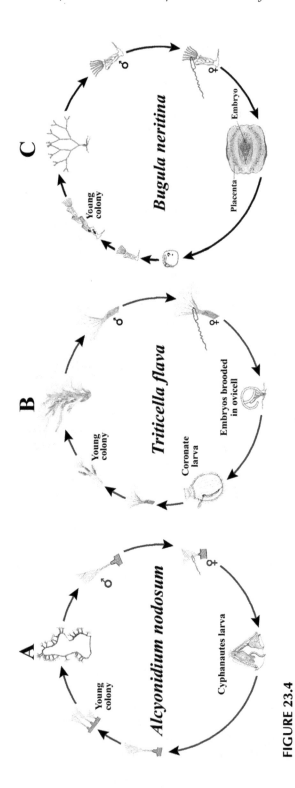

FIGURE 23.4

A. Indirect life cycle involving planktotrophic cyphanautes in gonochoric *Alcyonidium nodosum*, B. ovicell brooding and lecithotrophic coronate larva in *Triticella flava* and C. matrotrophic placentation and free-swimming larva (cooper@mail.uwlax.edu) in *Bugula naretina*.

account on the structure and functions of different ovicell types, Ostrovsky (2013) may be consulted. The egg size in these brooders ranges from 100 µm to 400 µm. These eggs develop into non-feeding lecithotrophic larvae. The indirect life cycle of cheilostome *Tendra zostericola* involves the coronate larva with non-functional rudimentary gut. The ctenostome larva of *Triticella flava* (Fig. 23.4B) looks like a cyphonautes but without a mouth and an anus. In the other ctenostomes *Flusterellidra hispida, Pherusella turbulosa* and *P. brevituba*, the larvae have retained an incomplete gut.

Pattern 3 is reported from phylactolaemates and some gymnolaemates. It is characterized by asynchronous development of numerically small oligolecithal oocytes in the former and sequential maturation of mesolecithal eggs in the latter. Intra-ovarian fertilization is followed by sequential brooding in the polypide (ctenostomes) or in the brood chamber in others. According to Ostrovksky (2013), the egg size in them ranges from as small as 50 µm in *Baenia bilaminiata* to as large as 300 µm in *Cribricellina cribraria*. Consequently, the gradation of matrotrophy also ranges from incipient to substantial and corresponding embryo enlargement. For example, it increases from 1.5 fold in *Figularia figularis* to 3.0–3.4 fold in *C. cribraria*, 53-fold in *Mollia multijuncta* and 310-fold in *Bugula neritina*; within the genus *Bugula*, it ranges from 6.3-fold to 500-fold in different species. The life cycle of matrotrophics, especially the viviparous epistomids is direct (Fig. 23.4C). The estimate on the number of placental matrotrophic bryozoan species is ~ 1,000 species, the highest among aquatic invertebrates. In bryozoa, matrotrophy independently evolved at least 22 times.

Fertilization and Development

Fertilization: The existence of protandrous zooidal hermaphroditism, the massive production of spermatozoa and their long lifetime and swimming capacity for a longer duration in the surrounding water led Joilet (1877) to propose that in bryozoans, cross fertilization is the rule. Subsequent studies have revealed that cross fertilization is indeed promoted by the long life span of sperm (Manriquez et al., 2001) and high efficiency of sperm capture by feeding lophophore (Pemberton et al., 2003). The spermatozoa of *Celleporella hyalina* remain viable for 2–3 weeks. Experiments with isolated colonies of *C. hyalina* showed that intracolonial self-fertilization in most cases resulted in abortion of embryos, reduced production and fitness of the larvae (Hughes et al., 2009). In isolated colonies of *Bugula stolonifera*, selfing produced viable larvae that could also complete metamorphosis successfully. But the colonies that were formed from selfing did neither survive long nor were fit. These experiments by many authors go to show that the cross fertilization may be the rule in bryozoans.

Embryonic development: Available information on embryonic development is summarized by Santagata (2015b). In broadcast spawning bryozoans, as exemplified by *Hislopia malayensis* (Wood, 2007) and *Membranipora membranacea* (Santagata, 2015b) that develop into planktotrophic cyphonautes larva (Hyman, 1959), the first cleavage is holoblastic, equal and pass through the Animal-Vegital (AV) axis (Fig. 23.5B). The second one is also holoblastic, equal and passes through the AV axis but rotates at 90° to that of the first cleavage plane (Fig. 23.5C). With this, the pigmented cytoplasmic granules are segregated to the vegetal cells. The third one is also equal and all the 8-cell, the pigmented granules are relegated to the 4-vegetal cells (Fig. 23.5D). The fourth cleavage passes through the AV axis resulting in a bilaterally symmetrical 16-cell embryo, composed of two rows of four equally sized cells in the animal pole stacked on the eight vegetal cells. The fifth cleavage produces 32-cell embryo comprised of 4 tiers of cells, consisting of 8, 12, 8 and 4 cells in the AV axis. Gastrulation commences with internalization of four vegetal cells, as the embryo approaches the 64-cell stage (Fig. 23.5G). The top animal tier of 8-cells develops into an apical disk. These cells ultimately form the ectodermal and endodermal tissues of the polypide. The second tier of 8-cells develops into hundreds of multi-nucleated coronal cells used for locomotion. In some non-feeding larval forms these cells undergo cleavage arrest of 32–40 cells in the coronal band. The third tier develops into larval pyreform complex, which is ultimately used for creeping and food capturing. The fourth tier contributes to the larval mesodermal tissue. On completion of the embryonic development, a cyphonautes larva (Fig. 23.5H) is released.

Cyclostome bryozoans brood their embryos in special enlarged gonozooids. Their fertilized eggs undergo cleavages that are considered

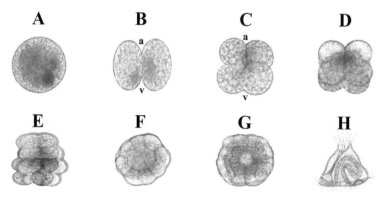

FIGURE 23.5

Embryonic development in *Hislopia malayensis*. A = egg, B, C, D and E represent the 1st, 2nd, 3rd and 4th cleavage stages, respectively (Wood, 2007); F, G represent the 5th cleavage stage and post-gastrula stage in *Membranipora membranacea* (Santagata, 2015b); H, the fully developed cyphonautes larva. a and v represent the animal-vegetal pole axis (modified and redrawn from Hyman, 1959, Wood, 2007, Santagata, 2015b).

to be irregular in cell numbers. The enlarged two-cell layered primary embryo generates other genetically identical secondary embryos through polyembryony (Pemberton et al., 2007). Limited information is available on early embryonic development of freshwater phylactolaemates, and is based on one species *Plumatella fungosa*. In it, the early cleavages are holoblastic and irregular, producing and elongated hollow blastula consisting of a single layer of cells. Following gastrulation, matrotropy provides maternal nutrients for further development through the tissue connection between the ovisac and the equatorial region of the embryo (Santagata, 2015b). Being sessile, the viviparous epistomiids also produce a free-swimming larva that re-establishes the colony in an adjacent area.

Regeneration and Clonal Multiplication

All bryozoans are capable of clonal multiplication by budding and fragmentation. In phylactolaemates, the bud primordium originates by proliferation of the basal (in vertical colonies but central in spherical colonies) undifferentiated cells of the epidermis, following the disappearance of muscular layers and basement membrane of the body wall at the site of proliferation. Thus, the basal epithelial cells, the descendants of the top animal tier of 8-cells, constitute a store of pluripotent embryonic stem cells that are transmitted from generation to generation (Braem, 1890, Marcus, 1934). In a simpler type such as *Fredericella* and *Plumatella*, buds appear in a definite order on a clear area of the maternal cystid wall ventral to the vestibular region. Each zooid bears two bud primordia: a main bud, which forms the first daughter zooid and an adventitious bud between the main bud and parent zooid. The adventitious bud becomes the second daughter zooid. In addition, each main zooid itself has a small duplicate bud primordium on its ventral side. In this budding process, duplicate adventitious buds become the main buds, and new duplicate and adventitious buds are formed (Wood, 2010). The process continues, producing a branching colony. A polypide can give rise to only a limited number of successive adventitious buds. As a result, the older, more basal/central polypides do not bud any longer and undergo disintegration. As the colony size increases, the elongated bilaterally symmetrical *Cristallata* colony fragments into smaller colonies. In spherical colonies, the daughter colonies are produced by fragmentation(s) at the periphery. In *Cupuladria exfragminis*, auto-fragmentation involves the development of distinctive morphologies designed to facilitate fragmentation. The growth rate in the colonies is significantly higher (7.5 µm/d) following auto-fragmentation than in fragments (5.6 µm/d) that originated from a mechanical disturbance (O'Dea, 2006). The indicated epidermal proliferations invaginate, carrying with it the peritoneum, and forms a two layered vesicle. Descriptions of further development by budding differ from author to author, perhaps due to species specific variations. For more details, Hyman (1959) may be consulted.

Except for some ctenostomes that produce hibernacula, gymnolaemates do not produce statoblasts, a 'seed-like' clonal product that facilitates the phylactolaemates to tide over unfavorable conditions. However, a striking feature of gymnolaemates is the constant progress of colonial rejuvenation by degeneration and regeneration. Expectedly, almost all parts of the gymnolaemates do possess the required stem cells to regenerate the colony. Table 23.3 summarizes regeneration and clonal multiplication in bryozoans.

TABLE 23.3

Regeneration and clonal multiplication in bryozoans

Species/Reference	Reported Observations
Gymnolaemata Cheilostomata	
Membranipora (Bronstein, 1938)	Cauterization of regeneration site evokes the formation of two buds, one proximal and the other distal to the wound. The distal one regenerates a polypide; but it inhibits the proximal one.
M. crustulenta (Borg, 1947)	The surviving overwintered polypide gives rise to new polypide.
Bugula neritina (Aymes, 1956)	Isolated rhizoids give rise to new budding zooids; isolated pieces of branches can develop stolons for attachment
Bugula (Abeloos, 1951)	Following amputation, any zooid can bud a single file of asymmetrical zooids; the symmetry is restored subsequently.
Ctenostomata	
Zoobotryon pellucidum (Zirpola, 1924)	The dormant pieces of stem from a degenerating colony remain alive and give rise to a stolen, from which new zooids are budded out.
Victorella pavida (Braem, 1951)	Surviving overwintered dormant buds generate zooids and stolons.
Paludicella articulata (Harmer, 1913)	The spindle-shaped winter bud hibernacula harbor polypide primordium, from which polypides arise.
Bulbella abscondita (Braem, 1951)	Submerged stolon gives rise to polypides under favorable conditions.
Stenolaemata Cyclostomata	
Crisia (Harmer, 1891a)	Stumps of dead colonies give rise to zooids. They can also generate new rhizoids. Old cystid can also form new orifice, from which a new polypide is regenerated.
(Harmer, 1891b)	A broken zoecium secretes a calcareous diaphragm across the opening. Subsequently, a polypide bud arises beneath the diaphragm. Rhizoids are always capable of giving rise to stolons, which bud out zooids. Old zooids are always capable of developing rhizoid that originates a branch.

For phylactolaemates, available information is limited to a few contributions during the 1920s–1940s on experimental amputations that may help to identify the location of stem cells. This description is a brief summary of the then available information from Hyman (1959). When the tentacular ends of *Stolella* are removed (Fig. 23.6A), the wound closes by contraction and some

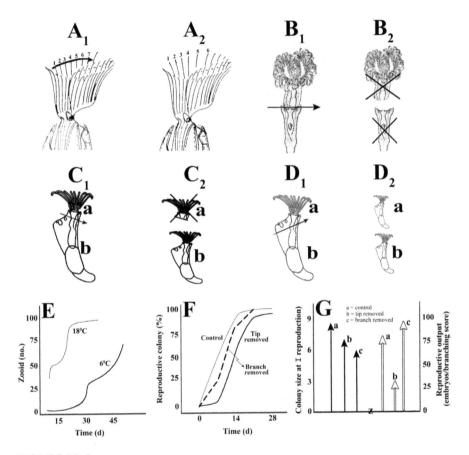

FIGURE 23.6

Stolella: (A₁–A₂) show the seven tentacles amputated and six tentacles regenerated lophophores. *Lophopus*: (B₁–B₂) amputated at the lophophore base and dead ramets. *Fredericella*: (C₁–C₂) amputation through the vestibular region. Anterior ramet without the budding zone dies but the posterior ramet containing budding zone and intact digestive loop regenerated the anterior and survives. (D₁ and D₂) amputation through the anterior ramet with budding zone and posterior ramet with intact digestive loop results in complete regeneration and survival of both ramets (drawn using reports by Marcus, 1934, Otto, 1921, Brien, 1936). E. Effect of temperature on regeneration rate, as measured by increase in the number of zooids *Membranipora membranacea* (modified and redrawn from Menon, 1972). F. Effect of tentacular tip or branch removal on proportion of reproductive zooids in *M. membranacea*. G. Effect of the same on size at first reproduction and reproductive output in *M. mebranacea* (modified, compiled and redrawn from Bone and Keough, 2005).

degree of dedifferentiation occurs in the adjacent cells to form a blastema. The tentacles grow from the stump but as a rule, fewer tentacles appear than those that have been removed. However, no regeneration follows, when the low lophophore arms of *Lophopus* are amputated near the bases or when the distal part of the polypide is removed at the esophagus or the ganglion level. Following amputation below the level of the stomach, the polypide degenerates within two days (Fig. 23.6B). Evidently, multipotent stem cells responsible for regeneration of the tentacles alone are located at the base of the lophophore. Apparently, *Stolella* and *Lophopus* rely on statoblast for clonal multiplication. However, when the anterior end of *Fredericella* is removed by a cut through the vestibular region without including the budding zone but with an intact digestive loop, the amputated anterior ramet dies; however, the remaining polypide regenerates a new anterior end (Fig. 23.6C). If the amputated anterior end includes the budding zone also, the remaining polypide with an intact digestive loop develops into complete polypide and the anterior end closes forming a small cystid with a degenerated lophopore, which regenerates a polypide at the site of attachment of the old esophagus (Fig. 23.6D). Apparently, the budding zone and digestive loop harbor pluripotent stem cells, responsible for posterior and anterior ramet regeneration, respectively. It is not clear whether some phylactolaemates like *Fredericella* possess pluripotent stem cells, as their potency for statoblasts production is limited. Information is available on growth and reproduction of regenerating gymnolaemates. For example, Menon (1972) described the effect of temperature on the rate of regeneration in *Membranipora membranacea* and *Electra pilosa*. Understandably, the rate is the fastest at 18°C and slowest at 6°C (Fig. 23.6 E). *M. membranacea* is intensely predated by the nudibranch *Polycera hedgpethi*. Bone and Keough (2005) reported the reproductive functions of the regenerating the *M. membranacea* colony following the removal of tentacular tip or branch (Fig. 23.6F). The branch-removed colony postponed the day of first reproduction but increased reproductive output more than the control unpredated colony (Fig. 23.6G).

Swimming larvae of bryozoans actively select attachment site. *Pectinatella magnifica* consistently avoid rock particles smaller than 1 mm diameter. *Paludicella articulata* settle selectively on mussel shells (Wood, 2010). Their success as space occupiers resides largely in their capacity for colonial growth by zooidal budding (Seed and Hughes, 1992). Encrusting bryozoans occur principally on macroalgae and some like *Membranipora isabellaenea* are obligate epiphytic, indicating that *M. isabelleanea* draw exudates arising from the fronds of *Lessonia trabeculata*. Grown on glass plates in the laboratory or in field for a month, the fronds-grown *M. isabellaenea* colonies retain a significantly higher proportion of live, non-degenerated zooids than colonies grown on glass plates. Obviously, the exudates influence the composition of live and degenerated zooids in a bryozoan colony. A second factor that influences the composition is the size and other features of the fronds (Manriquez and Cancino, 1996). Among the epiphytes, the red algal (*Odonthalia dentata, Phyllophora interrupta, Phycodrys rubens*) fronds persist for years in the White Sea, whereas the distal part of the kelp (e.g. *Saccharina lattisima*) together with epifauna may be torn off. However, the larger (4 m) and wider (0.2 m) kelp blade provides a larger substrate area than the red algae (0.3 m long, 0.05 m wide). This difference may impose changes in the composition of bryozoan zooids. For example, cheilostomid *Cribrilina annulata* colonies on the kelp are larger and have a higher proportion of dormant zooids, whereas the proportion of auto-, rejuvenating- and gono-zooids was higher in colonies inhabiting the red algae (Nekliudova et al., 2019). The life span of encrusting bryozoans range from transient species that grow rapidly, reproduce early and die (e.g. *Electra pilosa, Celleporella hyalina, M. membranacea*) to annual species with faster growth and earlier reproduction (*Flustrellidra hispida, Acyonidium* spp, Seed and Hughes, 1992) than those with a life span longer than 2 years (e.g. *Steginoporella* sp, Winston and Jackson, 1984, *Flustra foliacea*, Seed and Hughes, 1992). *Electra*, for example, can produce a chain of 'runners' to maximize the distance traversed across the substratum but with minimum attachment space; hence, it may prove to be a weaker spatial competitor on the fronds of *Fucus serratus* (O'Connor et al., 1980). Winston and Jackson (1984) describe the community development and life histories of encrusting cheilostomes, which are found abundantly under corals. A growth rate of 1.5 mm^2/d is required to achieve a longer life span and larger colonial size. It may enhance the chances of (i) colonizing new space, (ii) overgrowth and defense against competing neighboring sessile organisms and (iii) repair of injuries due to physical disturbance or predation. An apparent cost of such faster growth results in delayed sexual maturity; for example, the fast growing *Steginoporella* sp, *Reptadeonelle plagiopora* and *Stylopoma spongites* attain sexual maturity only at or after the age of 1.5 years. On the other hand, *Cribrilaria radiata* and *Parellisina curvirostris* grow slowly, mature earlier at the age of < 90 days and are recruited at a higher rate. Briefly, the cheilostome colony may grow fast and attain a large size but delay sexual maturity and thereby reduce recruitment rate. Alternatively, it may grow slowly, attain smaller colony size, mature earlier and recruited at higher rate.

Statoblasts and Hibernacula

Statoblasts: Phylactolaemates reproduce not only sexually and clonally but also by a special clonal method of producing statoblasts; Wood (2010) justifiably named them as encapsulated buds. According to the description by Hyman (1959), the statoblast is a cell mass enclosed within a protective shell that enables it to endure extremes of desiccation and temperature. Statoblasts are produced on the furniculus during the growing season following sexual reproduction. They are composed of cell masses arising from epidermis and furnicular peritoneum. The blastogenic epidermal cells migrate into the furniculus (Fig. 23.7A). The inwardly migrated blastogenic epidermal cells are then turned into a ball that soon becomes hollow to resemble a blastula (Fig. 23.7B), a feature of reproducing embryonic development. Such blastulae or vesicles cause a bulge in the peritoneal covering of the furniculus. While the vesicular cells are being flattened to a disk-shape, the enclosing peritoneal cells of the furniculus accumulate to its inner side and store reserved food in the form of yolk spherules. The two-layered epidermal disk grows around and encloses the mass of yolky peritoneal cells in a double epidermal wall. The inner blastogenic epidermal wall is destined to become the epidermis of the future zooid, while the outer wall secretes the statoblast shell. The thickened chitinous shell develops an equatorial suture between the dorsal or cystigenic and ventral or deutoplasmic valve (Fig. 23.7G). Then, the valves harden having a hollow interior that may be filled with air and thus a pneumatic ring or annulus is formed. Those statoblasts with well developed annulus of air cells (e.g. *Plumatella*, Fig. 23.7C) but without the armature of spines are called floatoblasts (e.g. *Hyalinella punctata*, Fig. 23.7D) and the floatoblasts bordered with hooked spines are named spinoblasts (e.g. *Cristallata mucedo*, Fig. 23.7F). Those statoblasts without or with a reduced annulus sink to the bottom or cemented to the maternal tissues are termed sessoblasts (e.g. *Fredericella sultana*, Fig. 23.7E). Interestingly, laboratory reared *Plumatella* colonies under optimum conditions frequently form both floatoblast and sessoblast (Wood, 2010). The number of statoblast formed by a polypide varies from one in *Fredericella* to a large number in *Plumatella*. For a colony, it may be numerous, for example, the number of statoblasts released by a colony of *Plumatella repens* runs to 800,000/m² of plant substratum (Brown, 1933).

Due to senescence, exposure to unsuitable conditions, following sexual reproduction or for want of food, the phylactolaemate polypides often degenerates. The degeneration commences with prolonged retraction of the polypide followed by disintegration in the sequence of the tentacle, interior tissues and the digestive tract. Under favorable conditions, statoblasts germinate; the germination involves the separation of two valves along the suture, appearance of germinal mass and projection of the polypide between the valves (Fig. 23.7H). From each statoblast, a single polypide named ancestrula (Wood, 2010) appears and originates in a manner similar to that in the clonal buds. The epidermal epithelium changes to a columnar form invaginating as the hollow vesicle, around which some yolky peritoneal cells gather to form a peritoneum. This two-layered primordium then proceeds to develop into a polypide. Hyman (1959) summarized the then available information on statoblast germination success following

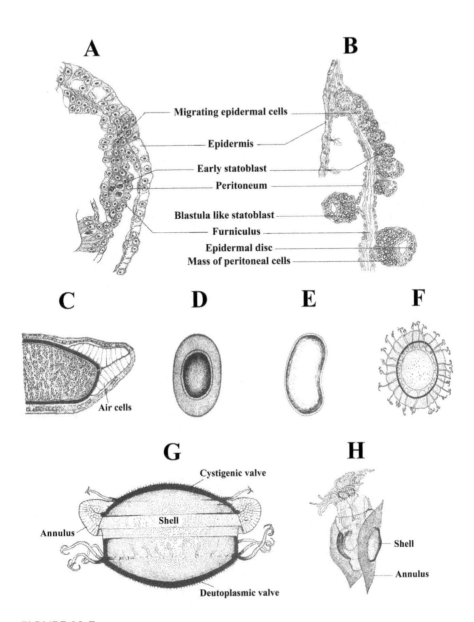

FIGURE 23.7

Formation of the statoblast in A. *Lophopus* and B. *Cristallata* (after Brien, 1953). Formation of capsule and annulus in C. *Plumatella* and D. floatoblast of *Hyalinella punctata* (after Brien, 1953, Rogick, 1940). E. Sessoblast of *Fredericella sultana* (after Allman, 1856). F. Spinoblast of *Cristatella mucedo* (after Toriumi, 1941). G. A section through empty statoblast of *C. mucedo*. H. Germinating *Lophopus crystallinus* polypide out of a statoblast (after Brien, 1953).

experimental exposure to desiccation. Expectedly, temperature is an important environmental factor that determines the success and duration of germination. For example, *Pectinatella* statoblasts germinated in 20 days at 11°C, 10 days at 15°C, 6 days at 19°C, 5 days at 23°C and 4 days at 27°C but died after beginning germination at 31°C. Hence, optimum temperature for *Pectinatella* lies between 19° and 27°C. *Lophopodella carteri* germinated after being kept at room temperature for periods of 1,543 d (4.5 years) but not after 6 years. The percentage of germination decreased with time of drying but with no change in germination time. *Fredericella sultana* statoblasts dried for 733–1,103 days germinated but no polypide developed. All statoblasts must remain dormant for a species-specific duration prior to their germination. *Plumatella casmiana* is unique in releasing thin walled subitaneous statoblast. The experiments of Brown (1933) on *Plumatella repens* and *P. magnifica* proved that statoblasts can germinate even without exposure to desiccation. In fact, even the repeated freezing for 300 days and thawing two to five times did not affect germination. Nevertheless, summer statoblasts removed from live colonies failed to withstand even a week of desiccation. Clearly, statoblasts within the degenerating colony are adapted to withstand desiccation or freezing. In general, desiccation is more detrimental than freezing.

In tropical Indian waters, Annandale (1911) reported the highest production of statoblasts during hot summer and statoblast germination at the onset of monsoon or cooler weather. The common species *Hyalinella punctata* flourish continuously producing sexual larvae and statoblasts throughout the year. In the temperate zone, statoblasts produced during early summer germinate in the same summer but those during late summer overwinter and germinate in the following spring. Interestingly, spring statoblasts of *Pectinatella* require 3–6 days to germinate but the summer ones > 60 days (Brown, 1933). Thus most temperate species have two cycles annually germinating from statoblasts. Due to the restricted growing season in the Arctic, sexual reproduction is limited to once a year. It may often be altogether omitted in northern climates or in Alpine lakes due to shorter growing season. In fact, the production of statoblasts by many plumatellids is so prolific that sexual reproduction seems a relatively unimportant means of recruitment (Wood, 2010). Whereas sexual reproduction usually lasts for 3–4 weeks in nature, clonal multiplication continues throughout the growing season in all colonies, irrespective of appearing from larvae or statoblasts.

Limited information is available on hibernacula, which are formed only by ctenostome species. The hibernacula are formed externally and resemble stolons or simple zooids, in which the wall is thickened. They are filled with a yolk-like material and are often attached flat to the substratum (Jung et al., 2017).

24

Brachiopoda

Introduction

The monophyletic (Santagata, 2015c) Brachiopoda or lamp shells are coelomate, sessile animals enclosed in bilaterally symmetrical but dissimilar (Fig. 24.1A, B) (but not in Linguliformea, Santagata, 2015c) valves attached directly (as in Craniformea) or by a stalk called pedicle or peduncle. The valves are dorsal and ventral in position, in contrast to the lateral positions in bivalve mollusks. They are hinged or articulate, as in Testicardines or unhinged or inarticulate, as in Escardines (Fig. 24.1C). The brachiopod body occupies the posterior third of the interior space in the mantle cavity between the valves and the mantle. The cavity is largely occupied by the voluminous lophophore, an anterior outgrowth of the body wall and varied in structure. Female lophophores are heavier than those of males (e.g. *Liothyrella uva*, Meidlinger et al., 1998). The brachiopod body is covered with a one-layered epidermis of varying in thickness in different regions and stratification over the nervous masses. In articulates, three sets of muscles, (i) the adductors or occlusors close the valves, (ii) the diductors or divaricators open the valves and (iii) the peduncular muscles. The well developed coelom is divided into an unpaired protocoel, and paired mesocoels and metacoels. The mesenteries consist of connective tissues clothed on both sides by a flattened peritoneum sparsely provided with cilia that keep the coelomic fluid in motion. The mouth, located centrally in the lophophore, leads to an esophagus, enlarged stomach with one or more voluminous digestive gland(s) and intestine, which terminates blindly in articulates (Fig. 24.1G) but opens by an anus into the mantle cavity in inarticulates. The open circulatory system consists of a mid-dorsal vessel located on the digestive tract and branches at both ends. One or a pair of metanephridia constitutes the excretory system. The nervous system consists of a small supraenteric ganglion and a subenteric ganglion, from which several minor lateral nerves give branches to the mantle, and adductor muscles (Hyman, 1959).

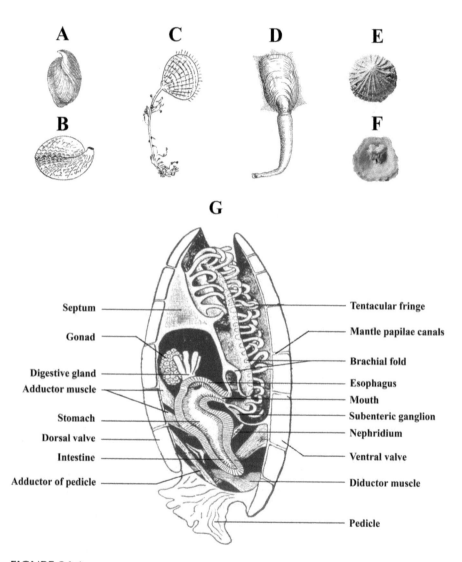

FIGURE 24.1

A. *Hemithiris psittacea* (after Blochmann, 1912), B. *Megellania* (Hyman, 1959), C. *Chlidonophora chuni* (after Chun, 1903), D. *Lingula* (after Blochmann, 1900), E. *Crania* (from revolvy.com) and F. *Novocrania* (from marinespecies.org). G. Shows anatomy of an articulate brachiopod (after Shipley, 1895).

Taxonomy and Distribution

From Hyman (1959), the classification of 391 speciose brachiopods has undergone many changes (Santagata, 2015c). However, the number of fossil records has remained around 30,000 described species. At present, the living brachiopods are divided into three subphyla: (i) Rhynchonelliformea (348 species; e.g. *Childonophora chuni*, Fig. 24.1C), (ii) Linguliformea (25 species; e.g. *Lingula*, Fig. 24.1D) and (iii) Craniformea (18 species; e.g. *Novocrania*, Fig. 24.1E, F). The most speciose rhynchonelliformids were earlier referred as articulate brachiopods. In them, the mineralized hinge connects the calcite valves. In the other two subphyla, the valves are held together only by muscles and connective tissues. The linguliformid shells are composed of apatite, a phosphatic mineral with an outer layer of chitin. Both rhynchonelliformids and linguliformids possess a peduncle; in the former, it is rigidly attached to a hard substratum but that of the latter is long and muscular structured and modified for burrowing into the soft sediments. With no pedicle, craniformids are cemented directly to hard substratum (Santagata, 2015c).

The Brachiopoda are exclusively marine inhabitants. They occur from the intertidal zone to depths of 5,000 m (Hyman, 1959). However, Schuchert (1911) estimated that 3% brachiopods live at the low-tide level, 81% on the continental shelves to the depths between 200 and 300 m, 3% in cold abyssal depths and the remaining 13% in the transitional zone. Known for their abundance in temperate and polar seas, the brachiopods are relatively sparse in tropical waters (Hyman, 1959).

Reproduction

Reproductive system: Most brachiopods are gonochores. However, simultaneous hermaphroditism is reported in the species belonging to the genus *Argyrotheca* (25 species, Hyman, 1959) and *Pumilus antiquatus* (James et al., 1992). *Fallax dallaniformes* and *Waltonia inconspicua* can be cyclic hermaphrodites (James et al., 1992). In brachiopods, gonads develop from mesolecithal folds of the peritoneum surrounding the gut. In Linguliformea, two pairs of gonads are developed from the mesenteries adjacent to the gut and on ripening, project into the metacoel. In others, they are developed inside the mantle coelomic canals positioned in the mesolecithal linings that underlie both shell valves (Santagata, 2015c). The gonads are unbranched L-shaped in Megathyridae (e.g. *Argyrotheca jansoni*), arborescent U-shaped in Terebratellidae (e.g. *Frenulina sanguinilenta*), reticulate in Laqueidae (e.g. *Laqueus californicus*), Hemithyridae (e.g. *Hemithyris psittacea*), Frieleidae (e.g. *Frieleia halli*) and Terebratulidae (e.g. *Terebratulina retusa*) or ovoid in Basiliolidae (e.g. *Neohynchia profunda*) (James et al., 1992). In the gonad, the germinal lamella, from which the gametes are developed, originates as part of the mesenteries supporting the stomach. In males, the lamella folds are ruffled to increase the surface area for sperm production, while in the female, a columnar construction is maintained. Both sperm and oocytes are generated on the same genital lamella in dorsal and ventral gonads. For more details on gametogenesis, James et al. (1992) and Meidlinger et al. (1998) may be consulted.

Spawning periodicity: In free spawners, the spawned gametes find their way to the exterior through the paired metanephridial gonoducts (e.g. *Lingula anatina, Terebratalia transversa*). But in brooders, the ova are retained in the lophophore or chamber, where they are cross fertilized in gonochores or may be selfed in hermaphrodites. Many brachiopods are known for longevity. For example, the longevity of gonochoric *Liothyrella neozelanica* is 30+ years and the lophophore brooder *L. uva* 46+ years (Baird et al., 2013). Spawning periodicity in these long living brachiopods may be once or twice a year or may be prolonged and asynchronous, as in *L. neozelanica* (Baird et al., 2013) or continuous, as in *L. uva* (Meidlinger et al., 1998). Within brooders, Chuang (1990) recognized four different adaptive strategies, of which the following may be mentioned: (i) some shed a portion of their ripe ova in a spawning and save the rest for a subsequent spawning in the same year, (ii) others like *L. uva*, despite spawning almost throughout the year in the Antarctic waters, stagger the release so that the brooded embryos are at different stages of development (Meidlinger et al., 1998) and (iii) still others, as like *Frenulina sanguinilenta* that produce 250 eggs but release and brood one at a time (Mano, 1960).

Fecundity: In a review, James et al. (1992) summarized the then available incomplete information on sexuality, development type, egg size and fecundity of 27 brachiopod species. In Table 24.1, the information is updated and completed as far as possible for 15 species. On rearranging it in three groups namely (i) free spawners, (ii) lophophore brooders and (iii) chamber brooders, the following became apparent. (1) *Sexuality*: Of 18 species, for which information is available, 12 are gonochores and 6 are hermaphrodites. Hence, a third of brachiopods may be hermaphrodites. (2) *In brooders*, the evolution of hermaphroditism became an adaptive strategy to ensure fertilization within lophophore or chamber. For, the lophophore brooders have to produce huge quantities of sperm which, following the release from males, have to travel a long way to fertilize ova retained in the female's mantle cavity. In fact, *Pumilus antiquatus*, listed as a hermaphrodite, has a sex ratio of 0.516 ♀ : 0.477 ♂ : 0.007 ♀̇, which may account for its lowest number of 75 embryos brooded in the lophophore. Brooding economizes the fecundity to 2 embryo/brood in *Gwynia capsula*, in comparison to the production of 60,000 egg/♀/y in *Glottidia* with external fertilization, which followed by embryonic development and an extended planktotrophic larval stage in the pelagic realm (Chuang, 1990). Notably, many brachiopod species that brood their embryos in a brood chamber are all hermaphrodites. (3) *Egg size*: Fecundity is counted by the number of eggs spawned by free spawners, but the number of embryos brooded in the lophophore or chamber. Within

TABLE 24.1

Sexuality, egg size, fecundity and spawning strategy in some brachiopods. G = gonochoric, H = hermaphrodite, L = lecithotrophic, P = planktotrophic (compiled from James et al., 1992 and others)

Species	Size (mm)	Sex	L/P	Egg (µm)	Fecundity
Free spawners					
Glottidia pyramidata	10 ?	G	P	90	60000
Lingula unguis	50	G	P	95	3000
Lingula anatina	60 ?	G	P	113	17250
Neocrania anomala	15	G	L	122	NA
Terebratulina retusa	18	G	L	140	11500
Terebratalia transversa	30	G	L	150	NA
Liothyrella neozelanica [††]	40	G	L	NA	NA
Lophophore brooders					
Liothyrella uva [†]	41	G	L	155	650
Terebratulina septentrionalis	20	G	L	160	NA
Terebratulina unguicula	25	G		170	NA
Notosaria nigricans	20	G	L	180	11340
Hemithiris psittacea	26	G	L	190	-
Pumilus antiquatus	5	H	L	200	75
Waltonia inconspicua	16	H	L	180	20000
Chamber brooders					
Lacazella mediterranea	10	H	L	20	150
Thecidellina barretti		H	L	20	150
Argyrotheca cuneata	5	H	L	95	25-80 oocytes
Argyrotheca cordata	5	H	L	100	

[†] = Meidlinger et al. (1998), [††] = Baird et al. (2013), *Argyrotheca* sp, Kaulfuss et al., 2013

gonochoric lophophore species brooders, the lecithotrophic egg size ranges between 155 µm and 190 µm but the hermaphroditic species are able to invest more resources in each egg, whose size ranges between 180 µm and 200 µm. In the brood chamber type, the egg size ranges from 20 µm in *Lacazella mediterranea* to 95–100 µm in *Argyrotheca* spp. In comparison to the egg size in lophophore brooders, it is not known whether these eggs are nourished by transfer of maternal nutrients. It is an established fact that with an increased egg size, fecundity decreases. However, on plotting the egg size of the brachiopods as a function of fecundity, a positive linear trend became apparent (Fig. 24.2A), but not among lophophore brooders (Fig. 24.2B). The reason for this seems to be that in the gonochoric lophophore brooder group, males may have to produce enormous numbers of sperm at the cost

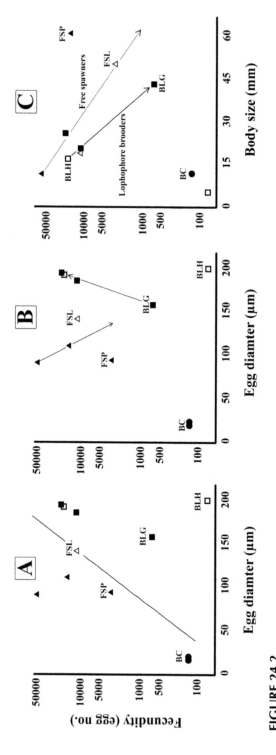

FIGURE 24.2

A. Fecundity as function of egg size and B. the same but as function of free spawners and lophophore brooders in some brachiopods. C. Fecundity as function of body size in free spawners and lophophore brooders. FSP = planktotrophic free spawners, FSL = lecithotrophic free spawners, BLG = lophophore brooding gonochores, BLH = lophophore brooding hermaphrodites, BC = chamber brooding hermaphrodites (drawn from data reported in Table 23.1).

of body growth to ensure fertilization in a relatively lesser number of eggs. Despite dual gonadal establishment, hermaphrodites grow to a larger size and still achieve a higher fertilization success. It is also an established fact that with increasing body size fecundity increases. Unexpectedly, the trends for fecundity-body size relation are all negative (Fig. 24.2C). Incidentally, it may be noted that its not clearly indicated whether the data reported in Table 24.1 are all for batch fecundity or seasonal fecundity. Nevertheless, further research is required to explain the discrepancy.

Embryonic Development

Cleavage: Typically, early cleavage is holoblastic, radial and regulated (Fig. 24.3A–D). Separated half- and one-quarter-blastomeres of *Hemithiris psittacea* and *Terebratulina septentrionalis* develop into smaller but normal larvae (Nielsen, 1991). Among brachiopod species, the cleavage pattern after the 4-cell stage differs and is not synchronized. In general, the animal half of the embryo is specified during blastula to a gastrula transition except in *Novocrania*, in which it occurs prior to fertilization (Freeman, 2000). In the vegetal half, the specification occurs at the 8-cell stage in *Glottidia* and *Novocrania* but later in the development of *Discinisca* (Freeman, 1999, 2000). In all of them, gastrulation, however, is initiated from the vegetal pole. The Antero-Posterior (A-P) axis is specified prior to cleavage in *Discinisca* and *Novocrania* but at the 16-cell stage in *Glottidia*. Left-Right (L-R) axes are also

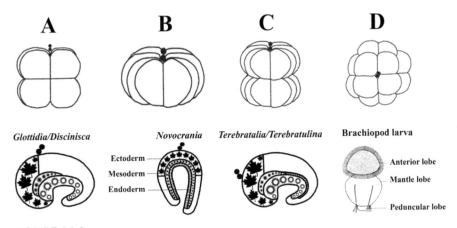

FIGURE 24.3

Variations in early cleavage stages of brachiopods. Upper row: 8-cell stalked radial unipolar arrangement in (A) *Terebratulina* and (B) *Terebratalia*, (C–D) 16-celled less common four tier stalking in *Novocrania* Lower row: Locations and distribution of ectodermal, mesodermal and endodermal primordia in late gastrula stage of different brachiopods (all are free hand drawings from Nielsen, 1991, Freeman, 1995, 1999, 2000).

specified at the same embryonic stage in all of them, except in *Novocrania* (Santagata, 2015c). These specifications have an implication for regeneration. In brachiopods, regeneration is limited to peduncle, involving the ectodermal descendent cells alone (Hyman, 1959).

Regarding tissue specification of embryonic germ layers, three types are recognized (Fig. 24.3). In *G. pyramidata, L. anatina* and *D. strigata*, as well as *Terebratalia*, mesoderm is formed from the anterior of the archenteron cells that ingress into the blastocoelic sites along the A-P axis. But the mesoderm arises from endoderm in *Neorhynchia anomala* and *T. transversa*. For more details on organogenesis, Chuang (1990) and Santagata (2015c) may be consulted. Finally, a free-swimming larva consisting of an anterior lobe, middle mantle lobe and posterior peduncular lobe is hatched.

Part F
Eucoelomates

Chaetognatha, Hemichordata and Urochordata constitute the eucoelomate minor phyla. Barring the speciose (3,000 species) urochordates, the others are not speciose, each may comprise a few hundred species. Hemichordata and Urochordata have been considered in Volume III of this series (Pandian, 2018). Chaetognaths are planktonic and are not capable of either regeneration or clonal multiplication. The others are benthic and have the potency for regeneration and clonal multiplication.

25

Chaetognatha

Introduction

The Chaetognatha are small (up to 100 mm but mostly ~ 40 mm), slender, torpedo-shaped, bilaterally symmetrical eucoelomate arrow worms with one (Fig. 25.1A) or two (Fig. 25.1B) pairs of horizontal lateral fins and a horizontal caudal fin. Their body is divided into a head, trunk and tail, and is covered with an extremely thin cuticle with an underlying one-layered epithelium and a thin basement membrane. A single layer of longitudinal muscle layer is located beneath the membrane. The peritoneal membrane is missing but the mesentery is covered by a thin layer of muscle fibers on both of its sides. The digestive tract is a straight tube commencing with a mouth located ventrally on the head and terminating into an anus (Fig. 25.1F). The spacious enterocoelous coelom is located between the digestive tract and body wall, and is divided into three compartments, which are separated by the septa one just behind the head and another behind the anus. A unique feature is the fold of body wall consisting of coelomic space that can be drawn over the head. The nervous system consists of cerebral ganglia connected by a pair of circumventral commeasures and a large ventral trunk ganglion.

Taxonomy and Distribution

In 1975, Reeve and Cosper indicated that Chaetognatha comprise ~ 75 species. Presently, there are 150 described species (Harzsch et al., 2015), i.e. the erection rate for chaetognatha is ~ 2 species/y. These 150 species are accommodated in 26 genera and in 3 families: (i) Spadellidae, (ii) Sagittidae and (iii) Eukrohniidae. The ever increasing number of molecular phylogenetic studies have shown that the chaetognaths are the most enigmative taxa for the following reasons: (1) Their ribosomal cluster is duplicated resulting in two classes of paralogous *18S rRNA* and *28S rRNA* genes, both of which are extremely diverse from other Metazoa. (2) The sequence of the intermediate

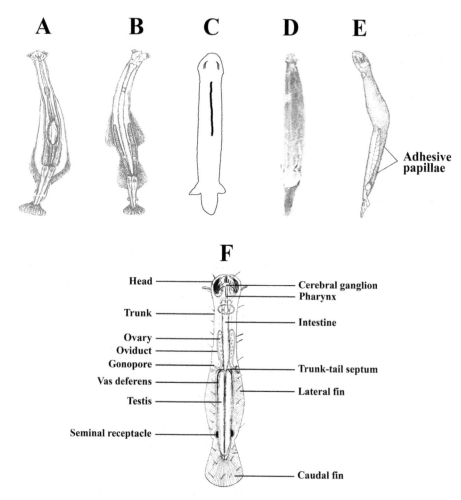

FIGURE 25.1

A. *Eukrohnia hamata* showing a single pair of lateral fins and B. *Parasagitta elegans* showing two pairs of lateral fins (redrawn from Grigor et al., 2017). C. *Paraspadella gotoi* (redrawn from http://evolution.unibas.ch). D. *Sagitta elegans* (drawn from an unknown source). E. *Spadella cephaloptera* (after Parry, 1944). F. Anatomy of *Spadella* (after Hertwig, 1880).

filament protein gene, for example, *Sagitta elegans* is very unusual compared to ~ 20 protostome and deuterosome sequences, suggesting a particularly high evolutionary drift of the chaetognath sequence. (3) The unique mosaic organization of the posterior/median *HOX* gene shows a long, isolated evolution of the chaetognath *HOX* complex. (4) Chaetognaths display a very unusually small mitochondrial size known for a metazoan and contain only 14 of the 37 usual genes (Harzsch et al., 2015).

The chaetognaths are exclusively marine and mostly planktonic. They occur in all oceans and at a wide range of depths, but no species are cosmopolitan in distribution. They are more abundant in tropical waters. The distribution of each species is limited by its reaction to depth, temperature (e.g. *Sagitta elegans* avoids warmer temperatures), light penetration, salinity (e.g. *Sa. bipunctata* in the Baltic Sea favors lower salinity) or combination of these factors. The mesoplanktonic chaetognaths like *Spadella lyra*, *Sa. maxima* occur in cooler waters below 200 m depth. However, some chaetognaths do occur at depths of 3,000 m, as in *Sa. serratodentata*, 5,500 m, as in *Sa. gazellae* and 8,000 m, as in *Sa. planctonis* (Hyman, 1959).

Reproduction

The chaetognaths are all protandric hermaphrodites. Their female reproductive system consists of a pair of solid elongated ovaries and each with its separate oviduct terminating in the vagina. It is located at the posterior part of the trunk (Fig. 25.1F). The male reproductive system also consists of a pair of band-like testes and each with its separate duct leading to a seminal vesicle, in which sperms are stored after copulation. The system is located at the anterior part of the tail. The two systems are separated by a septum. The ripe filiform sperm completely fills the caudal segment. In *Spadella*, they are packed into spermatophores. The sperm/spermatophores are released by the rupture of the seminal vesicle (Hyman, 1959, Reeve and Cosper, 1975). Though selfing is possible, cross fertilization seems to be the rule. A reciprocal mating process is described for *Sp. cephaloptera* and *Sa. setosa* (Reeve and Cosper, 1975). The fertilized eggs (each 0.2 mm) of *Sagitta* are released singly into sea water. *Spadella* lays a group of 12–16 eggs in a cluster at intervals of ~ 9 days almost throughout the year. As the eggs exit, the cement glands encircling the vagina secrete an adhesive coat and stalk the eggs, which are then attached to the sea weeds or other objects, on which *Spadella* oviposits (Hyman, 1959).

Chaetognaths are not amenable to rearing in a laboratory. Reeve (1970) was the first to rear and describe the direct life cycle from an egg to egg stage in a pelagic chaetognath *Sa. hispida*. Subsequently, Goto and Yoshida (1997) succeeded in rearing the benthic chaetognath *Paraspadella gotoi* and provided very useful information on the effect of feed and temperature on its survival and fecundity. They classified the growing arrow worm into four stages: (i) non-feeding, (ii) first feeding, (iii) second feeding and (iv) later stages (cf Alvarino, 1990). The first feeding stage was fed *Tigriopsus japonicus* nauplii, the second feeding stage either *T. japonicus* copepodites or newly hatched *Artemia* nauplii and the late stage of *P. gotoi* preferred *T. japonicus* adults. Each of the first and second feeding stage lasted for 10 days. Development from the first feeding stage to first oviposition at 3.5 mm size required ~ 50 days at 24°C but 70 days at 17°C. Survival was also strongly temperature-dependent, being higher at 17°C than at 24°C. Mature females lay a pair of egg clusters once every 12.6 day on *T. japonicus* diet at 17°C. Irrespective of temperatures at 17°C or 24°C, the cluster consisted of ~ 80 egg/cluster but ~ 9.5 egg/cluster, when fed on *Tigriopus* and *Artemia*, respectively. Food supply significantly affects oviposition; for example, mature *Sa. hispida* ceased reproductive activity on starvation (Reeve and Cosper, 1975). Interestingly, Reeve (1970) found 41% of the ingested nitrogen was utilized for gamete production in *Sa. hispida*. Clearly, quality and quantity of food supply plays a more important role than temperature in deciding the number of eggs to be packed in a cluster. Lifetime fecundity of *Sa. hispida* is ~ 100 egg/female, when reared at 17°C and fed *T. japonicus* (Goto and Yoshida, 1997). However, temperature seems to determine the life span of chaetognaths; for example, the span of *Sa. elegans* decreases from 206 days at 4.2°C to 91 days at 9.1°C (Sameoto, 1971). From a population study, Grigor et al. (2017) recognized *Eukrohnia hamata* as capital breeders with reserves stored in oil globules and *Parasagitta elegans* as an income breeder with no such reserves (cf Pandian, 2013). As a consequence, *E. hamata* spawns twice a year in spring and autumn but *Pa. elegans* spawn only in summer in the Canadian Arctic waters. However, in temperate Brazilian waters, *Pa. friderici* spawns almost throughout the year with a peak during spring, which is synchronized with a higher food supply and the raining season (Mendes et al., 2012).

Fortunately, information on the duration of spawning periodicity is available for chaetognaths distributed from Canadian Arctics (68°N) to Chesapeake Bay (37°N) in *Sa. elegans*, up to Miami (25°N) in *Sa. enflata*, and *Sa. neglecta*, *Sa. regularis* and *Sa. robusta* in the equatorial tropic (see Alvarino, 1990). On plotting the duration of reproductive period as a function of latitude (Fig. 25.2), it became apparent that (i) the waters between 23°N and 23°S mark the limit for chaetognaths to reproduce continuously almost throughout the year, (ii) beyond these limits, the reproductive duration is increasingly limited to a shorter duration and is synchronized with the bout of planktonic productivity and (iii) with increasing latitude, the reproductive duration is shifted from spring to summer and to early autumn, and (iv) with a short life span of 43 days, *Sa. elegans* may pass through five generations during the reproductive period at Plymouth but *Sa. gazellae* through a single generation in a year in the Arctics (see Alvarino, 1990).

Development

In chaetognaths, fertilization is internal and occurs in the ovaries prior to ovulation mediated by specialized somatic 'accessory fertilization cells'

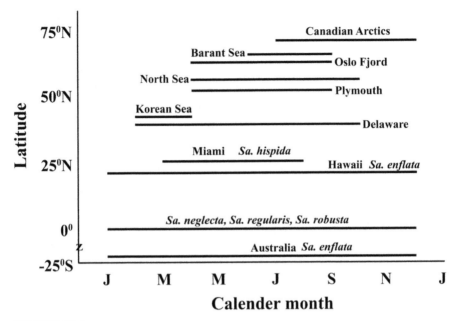

FIGURE 25.2

Duration of reproductive period in *Sagitta elegans* and others as function of latitude (drawn from data reported by Alvarino, 1990).

(Carre et al., 2002). The zygotes are then released into the water column in pelagic species (e.g. Sagittidae) or attached to plants or other objects (e.g. Spadellidae) or retained in a brood pouched for sometime (e.g. Eukrohniidae). Their development is direct and the hatchlings are almost similar to the adults in the structural organization.

Cleavage: The first cleavage is holoblastic and equal. At the second cleavage stage, the blastomeres are, however, displaced in such an alignment (e.g. *Paraspadella gotoi*) that development of the first four blastomeres has similarities to spiralians. Accordingly, they are designated as a, b, c and d blastomeres (Fig. 25.3A–H). An invagination of the blastopore initiates the endoderm formation and the stomodeal opening is then formed opposite to the blastopore. Hence, the developmental pattern of deuterostome is re-established, despite the initial variation to spiralian type. The endoderm folds inwards at two sites in the archenteron to form the mesoderm. The folds separate parts of archenteron opposite to the blastopore into three hollows. The middle hollow shrinks and eventually forms the intestine. The other hollows form parts of the head and trunk/tail paired coelomic cavities. With subsequent development, the mesodermal cells are rearranged into two bilateral groups and obliterated to form the cephalic and trunk coelomic

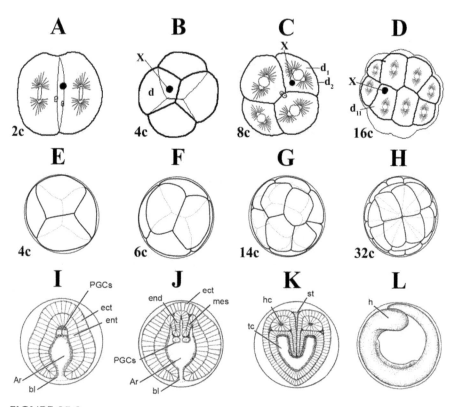

FIGURE 25.3

A–D. Early cleavage in *Sagitta bipunctata*. Note the lineage of X germ granule and d primordial germ cells (redrawn after Elpatiewsky, 1909). E–H. Early cleavage in *Spadella cephaloptera*. I–K. Gastrulation and mesoderm formation in *Sp. cephaloptera* (redrawn after Burfield, 1927). L. *Sp. cephaloptera* prior to hatching. Labels denote the 2-cell (2c), 4-cell (4c), 8-cell (8c) and so on stages. Ar = archenteron, bl = blastopore, ect = ectoderm, end = endoderm, h = head, hc = head coelom, mes = mesoderm, PGCs = primordial germ cells, st = stomodeum, tc = trunk coelom.

cavities (Fig. 25.3I–K). The d cell, destined to form the Primordial Germ Cells (PGCs), migrates to the mid-trunk at the level of the future posterior septum. Injecting single blastomeres of *Pa. gotoi* embryos with lineage dyes, Shimotori and Goto (2001) traced the fate of the four blastomeres. For details on organogenesis, Harzsch et al. (2015) may be consulted.

26

New Findings and Highlights

Introduction

As there are more books elucidating the phylogenetic relations among minor phyla, this book is largely devoted for the first time to aspects other than phylogenetics. The first finding is that with an average of 1,795 species/phylum, the minor phyletics are not as speciose as those (1,57,066/phylum) of major phyla. In sexual reproduction, recombination during meiosis and gamete fusion at fertilization produces genetic diversity —the raw material for evolution and its byproduct speciation. However, selfing hermaphroditism, parthenogenesis and clonal multiplication are some life history traits that deprive animals from producing genetic diversity. Hence, an attempt is made how these traits have reduced species diversity in minor phyletics.

Nevertheless, an introduction is required prior to the elucidation of hermaphroditism, parthenogenesis and clonals. With manifestation of sex a few billion years ago in organisms, sexuality has become an inherent trait of organisms and sexual reproduction perpetuates them. In some animals, it is supplemented with 'asexual' reproduction. There are two types of the so called 'asexual' reproduction. Progenies arising from clonal multiplication and parthenogenic reproduction tend to accumulate deleterious mutations and cause inbreeding depression; hence, many authors have considered them as 'asexual' reproduction. Yet, the differences between them must be recognized. The clonals appear from pluripotent or dedifferentiated stem cells but parthenogenesis from unreduced diploid eggs. The 'asexual' propagations exploit favorable conditions like abundant food supply and/ or optimum temperature by rapid multiplication of the fittest clone(s). With difficulty of finding a mate in patchily distributed populations, some have switched to hermaphroditism or parthenogenesis. With meiosis during gametogenesis, the hermaphrodites produce new genetic combination(s) but miss them at fertilization, as gametes arising from the same individual are fused. However, they have developed an array of strategies to avoid selfing. But parthenogens are deprived of producing any new gene combination, as neither meiosis nor fertilization occurs in them. Not surprisingly, parthenogenesis is chosen by < 1% of all animals (Bell, 1982), whereas 5–6% of them have opted for hermaphroditism (Jarne and Auld, 2006). Nevertheless, males do appear sporadically in parthenogens (except in bdelloid rofiers) and the rare occurrence of meiosis and fertilization may eliminate the accumulated deleterious mutations. In clonals, the probability of elimination of these mutations is minimal.

Hermaphroditism

In hermaphrodities, the manifestation and maintenance of dual sex within an individual may reduce fecundity to half of that in gonochores; correspondingly, the progeny number and hence, genetic diversity are also reduced. Motility enabling the search for a suitable mate is a decisively important requirement to manifest and maintain gonochorism. Surprisingly, gonochorism is established from the earliest minor phyla onward to Hemocoelomata, despite their low motility. Remarkably, transcriptome analysis has revealed molecular hints for the existence of spermatogenesis in Placozoa. In the absence of an organ system, the classical spermatogenesis is reported to occur in Mesozoa but within a single axioblast. Nevertheless, the combination of low motility and gonochorism has limited species diversity to < 1,000 species/phylum among the pseudocoelomates, except in Rotifera and Nematoda. With increasing structural complexity from Aorganomoprha to Acoelomorpha, Pseudocoelomata and to Hemocoelomata, almost cent percent gonochorism is maintained except in the acoelomate Gnathostomulida (Fig. 26.1). However, gonochorism begins to decrease from the lophophorate Entoprocta onward. Cent percent hermaphroditism occurs in Gnathostomulida (100 species), Bryozoa (5,700 species), Chaetognatha (150 species) and Urochordata (3,000 species). In lophophorates, some entoproct (18%) and phoronid (40%) species are hermaphrodites, irrespective of their colonial or solitary status. In all, (9,034 of 46,687 species) only 9034 species or 19.4% are hermaphrodites.

A vast majority (2,815 species, see Shenkar and Swalla, 2011) of urochordates are Simultaneous Hermaphrodites (SHs). Irrespective of commencing as males in protandrics or females in protogynics (e.g. *Botryllus schlosseri*, Cloney, 1990), they all terminate as hermaphrodites by simple addition of female or male components, as they grow in size or advance in age. This may also hold good for the protandric chaetognaths. As reciprocals, gnathostomulids act behaviorally as gonochores. The protandric chaetognaths are also reciprocals and may do the same as gnathostomulids. The selfers may exhibit one of the two evolutionary syndromes: a combination of high inbreeding depression

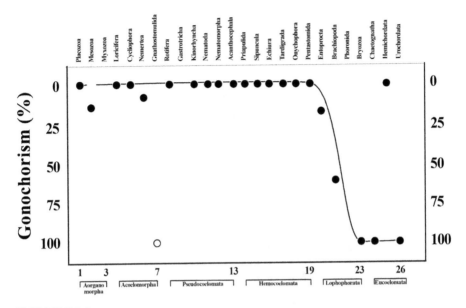

FIGURE 26.1

Percentage of gonochorism in minor phyla.

with low levels of selfing or low inbreeding depression with high levels of selfing. Selfing in the nemertean *Prosorhochomus americanus* belongs to the combination of low inbreeding depression with high level of selfing (see Pandian, 2021). In nematodes, a few are selfing hermaphrodites (see Pandian, 2021). In Tardigrada, some species in Macrobiotidae, Eohypsibiidae, Calohypsibiidae, Hypsibiidae and Necopinatidae are SHs. However, Nelson et al. (2010) reported that SH is limited to *Parhexapodibius pilatoi*, *Macrobiotus joannae*, *Bertolanius weglarskae* and four species in the genus *Isohypsibius*. Under laboratory conditions, Bertolani (2001) confirmed that isolated individuals in *I. minicus*, *M. joannae* were possibly selfers, as predicted by 'low-density model' proposed by Ghiselin (1969). Except for a few among Brachiopoda, almost all the hermaphrodites broadcast their sperm. This may also hold good for ascidians. When confined to a small disk, only 2% individuals of *Sagitta hispida* have been found to self (Reeve and Cosper, 1975). Dicyemid mesozoans (42 species) and a few ascidians (e.g. *Corella inflata*, *Ascidia callosa*, *Phallusia mamillata*, Cloney, 1990) are selfers. On the whole, ~ 100 species may be selfers. Briefly, the minor phyletics have neither opted for selfing SH nor for a costlier sex change. In them, the observed 19.4% SH may not be a reason for reduction in species diversity.

Parthenogenesis

The taxonomic distribution of parthenogenesis reveals its first appearance in acoelomate Loricifera, peaking in pseudocoelomate Rotifera, subsequent progressive decrease in Gastrotricha–Nematoda–Nematomorpha and hemocoelomate Sipuncula–Tardigrada and complete disappearance beyond hemocoelomates (Fig. 26.2). From Table 26.1,

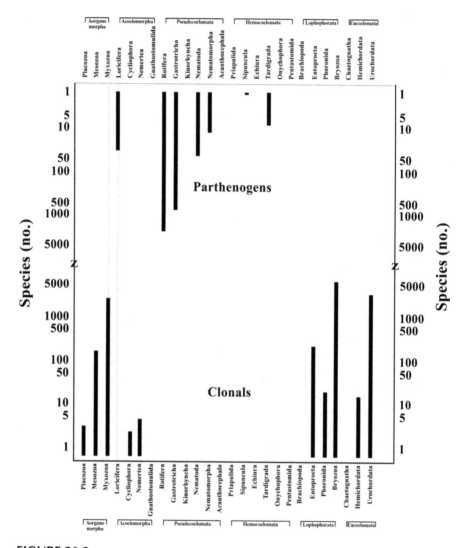

FIGURE 26.2

Approximate number of parthenogenic and clonal species in minor phyla.

TABLE 26.1

Approximate number of parthenogenic species in minor phyla. Mar = Mar, FW = Freshwater, T = Terrestrial, Par = Parthenogenesis

Phylum	Habitat	Mode	Species (no.)	Fecundity (egg no.)	Reference
Loricifera	Mar	Par ?	**34**	~ 12 ?	Kristensen (2002)
Rotifera	FW	Mitotic	**2028**	~ 22	Chapter 9
Monogononta			1520		
Bdelloidea			461		
Gastrotricha	Mar	?	**700**	~ 10	Hummon (1984)
	FW		**113**		
Nematoda					
Plant parasites	T	Mitotic	19		Liu et al. (2007)
		Meiotic	9		
Animal parasites	T	Meiotic	12		Table 12.3
Free-living	T	Mitotic	1		
		Meiotic	4		
			> 46		
Nematomoprha	FW		355		
Paragordius			11 of 18		Hanelt et al. (2012)
P. varius				2,950,000	Hanelt (2009)
Subtotal: 366	1% of 366		**> 11**		
Tardigrada					
Heterotardigrada	FW	Meiotic	247*		Bertolani (2001)
Macrobiotidae	FW + T	Meiotic	229*		
Hypsibiidae	FW + T	Meiotic	337*		
Milnesiidae	FW + T	Meiotic	15*		
Eohypsibiidae	FW + T	Meiotic	8*		
Subtotal	1% of 836		**8 ?**		
Marine			734		
FW			2743		
T			44		
FW + T			589		
Grand total			**2938**		

Values in bold letters represent the number of parthenogenic species. * = represent the number of species in respective family and not the number of parthenogenic species the following may be inferred: (i) With the higher level of structural organization, the lophophorate and eucoelomates opt for gonochorism or hermaphroditism but not for parthenogenesis. (ii) The relatively, structurally complex tardigrades opt for meiotic parthenogenesis. (iii) More than 98.6% of parthenogens are limited to freshwater, limno-terrestrial or terrestrial habitats. The existence of parthenogenesis in 34 speciose Loricifera and a single species Sipuncula (*Themiste lageniformis*) are more an exception. Without providing any evidence, Kristensen (2002) simply 'assumed' that the neotenic larva appearing from the Higgins larva of Loricifera commences the parthenogenic cycle. In *T. lageniformis*, male ratio ranges from 0.00 in Missouri Key to 0.04 in St. Lucie Inlet of Florida (Pilger, 1987). The ('protogynic') parthenogenic gastrotrichs terminate as hermaphrodites. Mitotic or apomictic parthenogenesis occurs in almost all rotifers and a few parthenogenic nematodes only, while most others are meiotic parthenogens (Table 12.3).

Whereas parthenogenesis seems to be a phylum-level feature of Rotifera and Gastrotricha, it is limited to a race (e.g. *Meloidogyne hapla*), population (e.g. *Pratylenchus scribneri*) or species-level in Nematoda and Tardigrada. Through a computer search, an assessment of the approximate number was possible for parthenogenic nematodes (Table 12.3). But, it was not possible for Nematomorpha and Tardigrada, as relevant information is not available. Table 26.1 shows that the approximate number of parthenogenic species may not exceed 3,000 in minor phyla, i.e. of 46,687 minor phyletic species, only 6.4% are parthenogenics. Among them, at least 25 nematode species are meiotic parthenogens; some tardigrades may also be meiotic parthenogens. Hence, only 6% of minor phyletics are mitotic parthenogens, which may suffer from inbreeding depression. In most of them, parthenogenesis is also supplemented with sporadic or regular (e.g. mono-, di- or poly-cyclic rotifers) events of sexual reproduction in rotifers and possibly in gastrotrichs. However, two points must be noted: (i) the bdelloids reproduce exclusively by apomictic/mitotic parthenogensis and yet have persisted and diversified into 461 species during the last 60 million years. In the 1,520 speciose monogononts too, only a few pass through monocyclic parthenogenesis, while most others are dicyclic or polycyclic (> 12 times, e.g. *Brachionus*, Gilbert, 2002); as a result, as many as 10^{12} amictic females can be generated within 60 days (Ruttner-Kolisko, 1963). The free-living nematode *Deladenus siricidicola* undergoes repeated parthenogenic cycles on pine trees and then undertake a single parthenogenic cycle as parasitic on *Sirex nocilio* (Fig. 12.12). With increasing scope for inbreeding depression, these parthenogenic minor phyletics have managed to survive over millions of year.

Clonals

Incidence of clonals covers a wider range of minor phyla from all the three aorganomorphic phyla (Placozoa, 3 species, Mesozoa, 150, Myxozoa, 2200) to Nemertea (4), Entoprocta (200), Phoronida (23) and Bryozoa (5,700) and to eucoelomates pterobranch hemichordates (~ 25) and ascidians (~ 1,800, i.e. 60% of all ascidians, Shenkar and Swalla, 2011). For the first time, it was reported that of 46,687 minor phyletic species, there are ~ 10,100 species or 21.6% are clonals and have the potency for clonal multiplication. Incidentally, that none of pseudocoelomate and hemocoelomate species clonally multiply confirms the hypothesis proposed by Murugesan et al. (2010). It is not known whether they do not have pluripotent stem cells, or are unable to dedifferentiate and to produce pluripotent stem cells to reproduce clonally.

Interestingly, a comparison on the incidences of parthenogenesis and clonals reveals mutual exclusion of each other (Fig. 26.2). A similar comparison has also disclosed that parthenogenesis and hermaphroditism also mutually eliminate each other (Fig. 26.3), except in some tardigrades. However, a look at Table 18.2 reveals that the mutual elimination also holds good at the family level for tardigrades. In terms of proportion of species too, clonality (21.6% of all minor phyletic species) and hermaphrodites (19.4%) predominate parthenogenesis (6.4%). In view of 1% incidence of parthenogenesis in the animal kingdom (Bell, 1982), the 6% parthenogenics in minor phyletics is still very high, which can be traced to eutelism in pseudocoelomates.

Not much information is yet available on proportions of sexual and clonal multiplication in Aorganomorpha. In the placozoan, *Trichoplax adhaerens*, only one egg is sexually produced per generation. But the clonal propagation by fragmentation increases the progeny number much faster. Whereas selfing hermaphroditism supplemented by clonal multiplication seems to have limited the species number to 42 in dicyemid mesozoans, gonochoric sexual reproduction alternated with clonal multiplication has enriched the species number to 108 in orthonectid mesozoans. The life cycle of Myxozoa is dominated by endogony, a rare mode of clonal multiplication in a single celled spore containing one to several hundred cells in its cytoplasm. Hence, cloning has perhaps reached the climax in the endogonic multiplication. Incorporation of a relatively more motile vertebrate (than the sessile bryozoa or sedentary oligochaete as definitive hosts) intermediate host, in which endogony occurs, has facilitated species diversity in myxosporeans. In Nemertea, only four species belonging to the genus *Lineus* are reported to have the potency for cloning. Recently, Ament-Velasquez et al. (2016) reported that polyploidy and hybridization has increased heterozygosity in *Lineus*, especially *L. pseudolacteus*.

Most lophophorates are sessile colonials, but others are solitary and may be tubiculous. Irrespective of gonochorism or hermaphroditism, they broadcast

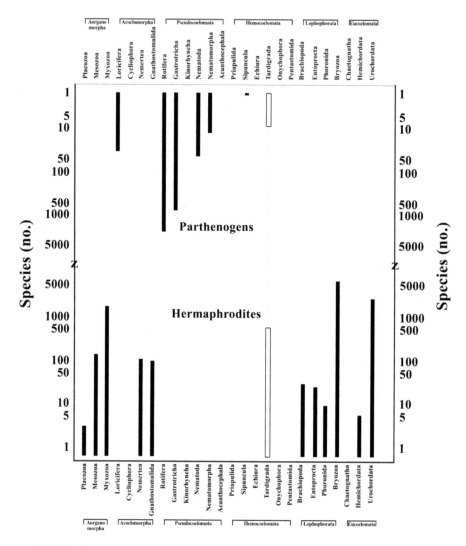

FIGURE 26.3

Approximate number of parthenogenic and hermaphroditic species in minor phyla.

their sperm but retain the eggs. The lophophores serve to capture con-specific sperm. With internal fertilization, many of them brood their embryos but broadcast the larvae for dispersal. With shells housing the animal, brachiopods are not cloners. In fact, the shelled animals like mollusks, ostracods and cirripedes are not cloners. Relatively more information is available *per se* for Entoprocta, Bryozoa and Urochordata. Japanese scientists have shown that irrespective of solitary or colonial status, the entoprocts are characterized by the year round predominance of clonal propagation and rare incidence of sexual multiplication. The clonal multiplication is so predominant that it has excluded sexual reproduction altogether in many species. For example, of 19 species in the genus *Loxosomella*, sexual reproduction is reported only in 9 species: *L. almuganecarensis*, *L. ameliae*, *L. cricketae*, *L. museriensis*, *L. pes*, *L. raja*, *L. studiosorum*, *L. tethyae* and *L. vivipara* (Sugiyama et al., 2010). The remaining 10 species reproduce almost solely by clonal multiplication. Even within some of these 'sexual' species like *L. plakorticola*, of 2,300 individuals examined over a period of two years, only 16 (0.67%) of them were found to reproduce sexually. In them too, the embryo number was limited to 5/individual, in comparison to equal number of individuals simultaneously reproduced clonally (Sugiyama et al., 2010). The clonal reproduction was not only predominant in the solitary *Loxosomella* but also in the colonial entoproct *Barentsia discreta*, in which the frequency of sexual reproduction, as evidenced by the presence of embryos in the lophophore, is limited to 0.2% at low temperatures and 38.6% at higher temperatures (30°C) (Iseto et al., 2007). On the whole, the species number for Entoprocta is limited to 200 due to the predominance of clonal propagation over sexual reproduction.

In bryozoans, three reproductive patterns can be recognized: (1) Annuals with LS of 3–12 months; they commence the first sexual reproduction at an earlier age and thereby provide a small proportion of clonal multiplication. (2) Perennials with LS of > 1 year; they delay the age at first sexual reproduction and thereby provide a relatively larger proportion of clonal multiplication. (3) The short lived plumatellids are characterized by clonal multiplication and sexual reproduction. In these clonals, there are two types: (a) the usual budding/fragmentation and (b) statoblasts production. Briefly, sexual reproduction is supplemented by clonal multiplication in pattern 1 of annuals. But clonal multiplication is supplemented by sexual reproduction in pattern 2 of perennials. In pattern 3, the plumatellids are characterized by predominant clonal multiplication with rare sexual reproduction. As a result, the dominance of clonal reproduction has limited the species number to 96 in the freshwater plumatellids. The 4,921 speciose gymnolaematids include patterns 1 and 2. As seven of 15 coral encrusting bryozoans belong to the pattern 1, it is likely that the patterns 1 and 2 approximately share 50% species each among the remaining eight species.

For bryozoans, the life span of some polypides ranges from 6 to 72 days (Gordon, 1987). The ovary is formed only once in most species or twice in a few others. The cheilostome broadcast spawners produce 4–5 to 40–50 small oligolecithal eggs per zooid at a given time. However, oocytes are produced over a period of several weeks by *Electra pilosa* and *Membranipora membranacea* indicating that their oogenesis is continued after the spawning of ovulated eggs and that this process is repeated several times. Yet, available information is scattered and reported in different units and formats. This account has attempted to assemble and 'process' them into a single format. Menon (1972) reared three marine species at different temperatures for short (60 days) and longer (120 days) durations (Table 26.2). From the growth

TABLE 26.2

Growth by addition of auto- and gono-zooids in some bryozoans

Species/Zooid Type	Short Duration	Long Duration
Reared at 18C for 60 or 120 days (Menon, 1972)		
Membranipora membranacea	2 zooid/d	Pattern 1
Conopium reticulatum	67 zooid/d	50 zooid/d, Pattern 1
Electra pilosa	20 zooid/d	42 zooid/d, Pattern 1
Celleporella hyalina **reared at unrestricted food supply for 120 or 180 days (Hughes, 1989)**		
Autozooids	14.3 zooid/d	11.7 zooid/d, Pattern 2
Male gonozooid	7.3 zooid/d	3.4 zooid/d
Female gonozooid	7.0 zooid/d	3.5 zooid/d
(Hayward and Ryland, 1973)		
Alcyonidium hirsutum	0.2 gonozooid/mm^2, irrespective of increasing colony size from 50 to 330 mm^2 area, Pattern 2	
Hippothoa sp	Number of female gonozooid linearly increases to 50 at colony area of 25 mm^2 Female gonozooids increases from 0.7 to 1.1 with increasing colony size, Pattern 1	
From field (Winston and Jackson, 1984)		
Gibrilauia radiata	Pattern 1	
Parellisina curvirostris		
Cleidochasma porcellanum	Pattern 2	
Typostega venusta		

trends of these bryozoans, the number of zooids added per day at 18°C was calculated. Hughes (1989) reared *Celleporella hyalina* under unrestricted food supply for short (120 days) and longer (180 days) durations. He also reported the number of autozooids and gonozooids. From the reported values, the production rate of autozooids, female and male gonozooids per day was estimated. In *Alcyonidium hirsutum*, Hayward and Ryland (1973) found the number of female gonozooids remained constant at 0.2 throughout the colony size up to 330 mm^2 area. But for *Hippothoa* sp, they reported a linear increase in the number of female gonozooids with increasing colony size up to 25 mm^2 area. However, they also included information on the increasing number of zooids with increasing colony area up to 13.5 mm^2. From these values, the number of female gonozooid was calculated to increase from 0.07 gonozooid/ autozooid at the colony size of 3.5 mm^2 to 1.1 gonozooid/autozooid at the size of 50 mm^2. Winston and Jackson (1984) published valuable information on survival, growth and reproduction in coral reef encrusting chilostome species. At moderately faster growth rate of > 1.5 mm^2/d, 7 of 15 species survived longer, grew larger but delayed the age at first sexual reproduction, for example, to > 135 days, i.e. (Life Span [LS] of 164 days, hence, Generation Time [GT] =) 82% GT in *Cleidochasma porcellanum* and to 450 days or 62% GT in *Typostega venusta*. By delaying the age at first sexual reproduction, these bryozoans provide considerable scope for clonal multiplication *per se*. On the other hand, in Pattern 1 with short LS (100–150 days) and GT, *Gibrilauia radiata* and *Parellisina curvirostris* began to sexually reproduce at the age of < 90 days. The fact that with increasing colonial growth, the number of (i) gonozooid increases in *Hippothoa* sp, (ii) autozooids remain at a low level in *M. membranacea* and (iii) autozooids decreases in *Conopeum reticulatum* suggest that sexual reproduction dominates in them. Hence, they may be placed in Pattern 1. But that the number of gonozooids (i) remains constant in *A. hirsutum* and (ii) decreases in *C. hyalina per se* indicates that clonal multiplication may dominate in them with increasing colony size. Hence, they are assigned to Pattern 2.

The freshwater plumatellid bryozoans (94 species) reproduce predominantly by clonal multiplication supplemented by sexual reproduction. Besides, they also clonally produce statoblasts, which perpetuate them, by tiding over the period(s), when the aquatic body is dried or frozen. There are two types of statoblasts: (a) The non-dispersible sessoblasts cemented to the substratum and (b) the dispersible floatoblasts. Some of these plumatellids are short lived annuals, for example, the LS of *Plumatella emarginata* averages to 1.6 months. *P. elongata* produce sessoblasts at the rate of 7.6/colony or 0.3/autozooid (Karlson, 1991). In it, the number of sessoblasts increased from 1 in a colony with 5.6 zooids to 38 in a colony with 169.8 zooids; these values are 0.1 and 28.7 for the floatoblasts (Callaghan and Karlson, 2002). In *P. ripens*, these values are 0.15/zooid and 0.50/zooid for the sesso- and floato-blasts, respectively. Karlson (1991) reported a linear positive increase in the relationship between the number of floatoblasts and sessoblasts on one hand and ramets containing increasing number of zooids on the other. With increasing number of zooids, the number increased up to 180 for the flatoblast but only up to 18 for the sessoblast. However, the former were produced only up to 120 zooid containing ramets but the latter up to 180 zooid containing ramets. It is likely that the production of sessoblast is costlier.

Table 26.3 provides available information on some life history traits and fecundity of hemichordates. With male ratio reduced to 0.02–0.08, finding a mate and acquiring adequate insemination may have limited the species number to 130 in hemichordates. With larger egg size of 1 mm, the heavier tornaria larva of the abyssal ptychoderids may have found it difficult to climb a vertical distance of > 3,000 m (Pandian, 2021). In the pterobranchs, the female ratio ranging from 0.01 to 0.09 and male ratio of 0.08 indicate that sexual reproduction may be a rare event. In them, clonal reproduction may dominate. As a consequence of inbreeding depression, not only is the species number is reduced to < 35 species but also their size.

For ascidians, literature available on sexual and clonal reproduction is extremely large. In terms of morphology and life cycle, the dichotomic types

TABLE 26.3

Life history traits and fecundity of hemichordates (compiled from Pandian, 2018). If = Internal fertilization, Ef = External fertilization, Dd = Direct development, Id = Indirect development

Taxonomy Group	Sex Ratio	Egg Number & Size	Fertilization/ Life Cycle
Hemichordata (~ 130 species, gonochores)			
Enteropneusta (< 100 species, solitary, burrowing)			
Harrimaniids	0.98 ♀ : 0.02 ♂	2,000 eggs, 1 mm	Ef, Dd
Ptychoderids	-	200–1,000 eggs 300 μm to 1 mm; 20,000 eggs 200 μm	Ef, Id
Pterobranchia (25 species, sessile colonials)			
Cephalodiscids	Many neuters	-	If
Rhabdopleurids	0.01–0.09 ♀ : 0.08 ♂ : 0.83–0.90 neuters	-	If

TABLE 26.4

Contrasting traits of solitary and colonial ascidians (modified from Tarjuelo and Turon, 2004)

Traits	Solitary	Colonial
Spawning	Broadcast	Brooding
Fertilization	External	Internal
Embryonic duration	Short	Longer
Larval size	Small	Large
Larval life	Longer	Short
Metamorphosis	Prolonged	Rapid

namely (i) solitary and non-cloner, and (ii) colonial and cloner forms have many implications to reproductive biology (Table 26.4). Many colonial clonal ascidians have developed viviparity and thereby produce large tadpoles from smaller eggs. As a result, for example, egg production commences in June at the colony size of 35 mm² area, peaks in December at 50 mm² size but the tadpole production peaks in March at 75 mm² size in the temperate *Didemnum rodriguesi* (Fig. 26.4A). Incidentally, the fecundity number per zooid and per colony linearly grows with increasing cumulative fecundity in *Ectenascidia turbinata* (Fig. 26.4B).

The phylum Urochordata/Asciacea comprises three orders: (i) Aplousobranchia with 1,480 species in 13 families, (ii) Stolidobranchia with 1,020 species in four families and (iii) Phlebobranchia with 315 species in nine families. As early as in 1935, Berrill described clonal multiplication by budding in 30 species from 21 genera and 7 families, indicating that

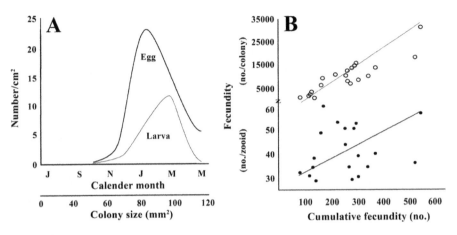

FIGURE 26.4

A. Egg and larva production in *Didemnum rodriguesi* as a function of colony size and calendar month (compiled from Ritzmann et al., 2009). B. Fecundity per zooid and per colony as a function of cumulative fecundity in *Ectenascidia turbinata* (drawn from the data reported by Garcia-Cagida et al., 2005).

the clonal incidence is widespread in all three orders (e.g. Clavellidae in Aplousobranchia, Stylidae in Stolidobranchia and Perophoridae in Phlebobranchia). Recent estimates on the incidence of colonialism and hence clonality (cf Blackstone and Jasker, 2003) reveal that the most speciose Aplousobranchia are clonals. For example, 14 of 14 genera, i.e. 100% of all the known genera among the most speciose aplousobranchids are colonials and clonals. This value decreases to 45 and 35% in the less speciose stolidobranchids and the least speciose phelobobranchids (Pandian, 2018). Of 64 documented non-indigenous ascidian species, 33 are colonials, in which 19 (57.5%), 12 (36.4%) and 2 (6.1%) belong to the aplousobranchids, stolidobranchids and phelobobranchids, respectively (see Shenkar and Swalla, 2011). In the former, the most speciose family is Didemnidae with 563 species and the genus *Didemnum* alone comprises 200 species (Shenkar and Swalla, 2011). And the *Didemnum* species are colonials and clonals (see Pandian, 2018). Therefore, some unknown factor(s) other than inbreeding depression may regulate species diversity among ascidian clonals.

Pandian (2018) identified that body size and reproductive potential as a factor that may regulate species diversity in clonal ascidians. For example, the repeated and rapid process of budding results in size reduction. In ascidians, the size ranges from 5–15 cm height in solitary non-clonal ascidians, in comparison to 5–10 mm width in the clonal ascidians. The life history traits of ascidians are one of the most diverse. It ranges from semelparity to iteroparity, oviparity to ovoviviparity, seasonal spawners (e.g. *Halocynthia roretzi*) to continuous breeders (e.g. *Ciona intestinalis*, *Oikopleura dioica*), and indirect development to direct development in the so called anurans. Data for fecundity are also provided in the number of egg/ zooid or larva/colony (e.g. Durante and Sebens, 1994) or number of mature oocyte + released larvae/zooid and colony, e.g. (Garcia-Cagide, 2005) and in terms of resource allocation for reproduction in units of weight and caloric content (Tarjuedo and Turon, 2004). To complicate it further, the released tadpole may have undergone partial or complete development (Tarjuedo and Turon, 2004). Table 26.5 lists the processed data for fecundity of clonal and non-clonal ascidians. Briefly, the fecundity of non-cloners ranges from 3,083 eggs (El-Sayed et al., 2018) to 1 million in *Phallusia mammillata*. But that for cloners, the range is 5/zooid in *Botryllus schlosseri* to 2,012 mature oocyte + 107 larva/colony in *Ectenascidia turbinata*, i.e. the fecundity of cloners is < 0.2% of that of non-cloners. These values show that with dominance of clonal multiplication, sexual reproduction is greatly reduced. But the question is: with clonal multiplication and consequent inbreeding depression, how does the species diversity of ascidians remain is so high?

Two features namely fusion and rejuvenization may be the reasons for the observed unusual species diversity in colonial ascidians, which clonally multiply by budding, fission and/or fragmentation. Several authors have reported valuable information on fission and/or fragmentation but only a few on fusions. Some of them considered fusion as a complicating nuisance

TABLE 26.5

Fecundity in sexually and clonally reproducing ascidians (compiled from Lemaire, 2011, Svane, 1983,** Garcia-Cagida et al., 2005,* Ritzmann, et al., 2009,[†] El-Sayed et al., 2018)[††]

Species	Fecundity (egg no.)	Life Span	Remarks	
Non-cloners			Egg (µm)	
Ciona intestinalis	10,000	2–3 mon	140	Semelparous
C. savignyi	10,000	2–3 mon	160	Semelparous
Phallusia mammillata	1,000,000	< 1 y	120	
Molgula oculata	5,000	1 y	100	
Halocynthia roretzi	30,000	3 y	280	Iteroparous
*Ascidia mentula***	> 100,000	~ 8 y	200	Iteroparous
*Pyura tossellata***	< 100,000	10 y	200	Iteroparous
*Boltenia echinata***	< 100,000		200	Ovoviviparous
Herdmania momus[††]	3,083			
Clonals				
Botryllus schollseri	5/zooid	< 3 mon		
*Oikoplauera dioica***	150	4 d		
*Ectenascidia turbinata**	60/zooid, 2012 + 107/colony			
Didemnum rodriguesi[†]	50/m³			

in estimation of colonial growth, while a few others reported it inadvertently. These few have also not looked at it from the angle of 'gamete' fusion at fertilization and thereby providing genetic diversity. Yes, the fusion of two fragments or colonies may represent 'fertilization-like event of two mega-gametes' to produce at least half the new gene combinations, as that in sexual reproduction, in which genetic diversity may be produced by (i) recombination during meiosis and (ii) fusion of gametes at fertilization. In didemid ascidians, colonies fuse indiscriminately (Bishop and Sommerfeldt, 1999), whereas in botryllids, fusion is regulated by genetic histocompatibility (Oka and Watanabe, 1957). Incidentally, Didemnidae is the most speciose family. Saito and Watanabe (1982) must be complimented for a clear description of the events in fusion in *Botryllus scalaris*. When the fusible colonies or their fragments come in contact (incidentally, colonial movement may promote fusion, Carlisle, 1961) with each other, the ampullae in the test of each partner at their growing/regenerating edges begin to extend toward each other (stage 1). Eventually, the boundaries between the two disappear, resulting in the fusion of the tests. With further growth and migration of ampullae of the partners towards each other (stage 2) contacts between the internal vessels are established (stage 3). At the contact points, the vessels are connected (stage 4). Subsequently, the fused ampullae are multiplied and transformed into internal vessels. Thus, the fusion is completed (stage 5). However, if the fusion is incompatible, the partners may separate at stage 4 or earlier.

Carlisle (1961) observed the occurrence of true fusion in *Trididemnum tererum*. Colonies can fuse at a wide range of size from the smallest of 0.9 mm^2 area to the largest of 116 mm^2; following fusion of the smallest colonies (12.1 mm^2), the fused colony measures 38 mm^2, perhaps due to immediate accelerated growth following fusion (see also Fig. 26.5B) (Stocker, 1991). During their investigation, Bak et al. (1981) found the disappearance of 43 colonies due to fusion in *Trididemnum solidum*. The fusion rate was almost double the times more (0.7/colony/d) in the presence of a sponge serving as substratum than that (0.4/colony/d) in the absence of it (Stocker and Underwood, 1991). In this ascidian, it increased from 0.05 time/colony/d in a small colony of 0.5 mm^2 size to 0.4 time/colony/d in a large colony of 20 mm^2 (Fig. 26.5A). The trend for the fission rate and colony size ran parallel to that of fission-colony size, suggesting that fusion between fragments occurred mostly after fission. However, colonial fusion can also occur even in the absence of fission (see Fig. 26.5B). Lopez-Logentil et al. (2005) found the fusion as a mechanism, by which some didemnid achieve the minimum size required to commence sexual reproduction (e.g. *Cytodytes dellechiajei*). In 51% of *Didemnum fulgens* colonies surveyed by Lopez-Logentil (2005), the highest incidence of fusion occurred during March–April at a time, when gametogenesis has begun (Fig. 26.5B).

As in other colonials, degeneration of colonies following rejuvenization by regeneration is not uncommon among ascidians (Ritzmann et al., 2009). For

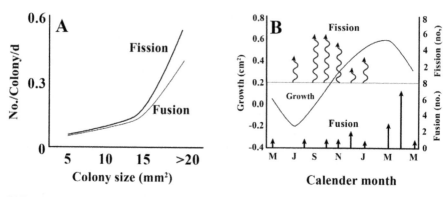

FIGURE 26.5

A. Daily incidence of fission and fusion as a function of colony size *Didemnum moseleyi* (modified and compiled from Stocker, 1991). B. Fission and fusion as a function of calender month in *D. fulgens*. Note the trend for colonial growth drawn in thin line (modified and redrawn from Lopez-Legentil et al., 2013).

example, proportion of the whole colonial death decreases with increasing colonial size in *Didemnum moseleyi*. In general, circular colonies degenerate at a slower rate than that of elongate ones (Stocker, 1991). Not all the dead colonies regenerate, only a few, perhaps the fittest one alone can regenerate. In this process, those suffering from inbreeding depression may be weeded out. However, not much information is yet available on this aspect.

Fecundity

It may be a good idea to remember the definition fecundity. Batch Fecundity (BF) is the number of eggs in a clutch or batch. Seasonal Fecundity (SF) is the number of BF multiplied by the number of spawning/oviposition during a breeding season. Lifetime Fecundity (LF) is the total number of eggs produced during the lifetime of an animal (see Pandian, 2011). Fecundity is a decisively an important factor in recruitment, and sustenance of population and species. Hence, an attempt is made to know whether a low fecundity limits species diversity in minor phyletics. Available information on fecundity of minor phyletics is limited to one or two publications (e.g. Placozoa) and is also reported in heterogenous formats and units; for example, the number of oocytes (e.g. Echiura), eggs or embryos (e.g. Bryozoa, Ascidiacea) and for batch (e.g. Entoprocta), or seasonal fecundity (e.g. Bryozoa) and for a colony (e.g. Bryozoa). In some of them, > 50% of the oviposited eggs are sterile (e.g. free-living nematodes) or unfertilized mictic eggs (e.g. Rotifera). Interestingly, entoprocts produce the maximum of five eggs per zooid each for sexual and clonal reproduction. In another entoproct *Cupuladria exfragminis*, auto-fragmentation produces 1.5 fragment/mo (O'Dea, 2006). In brachiopod, 5,000, 2,500 and 500 eggs are produced by free spawners, lophophore brooders and chamber brooders, respectively. Irrespective of all these heterogenous formats and units of the reported data, fecundity is limited to one egg in Placozoa to ~ 40 eggs in many Pseudocoelomata (e.g. Rotifera), except in the free-living nematodes with ~ 300 eggs. Irrespective of BF, SF or LF, the reported fecundity values do not exceed 300 eggs in Aorganomorpha to Hemocoelomata. It is only from the eucoelomate hemichordates, the LF is increased to > 1,000 eggs and steeply to > 50,000 eggs in ascidians.

An important reason for the low fecundity in lower minor phyletics, especially in pseudocolomates is the eutelism, i.e. the number of somatic cells remaining constant after hatching. The cell number is a few, about 50 in Mesozoa (Table 26.6), 430 in bdelloid and 790 in monogonont rotifers and 960 in nematodes. In them, the number of tissue types also remains low at 4–7 in Aorganomorpha, 15 (3–4 types of epidermal/gland tissue types, 3 nervous tissue types, 4 muscle tissue types and 4 gut tissue types) in rotifers and > 5 in nematodes. It is the somatic cells and tissue types that provide adequate nutrients for gametogenesis, development and maturation of gametes. Understandably, the low number of somatic cells and tissue types may play a decisively important role in limiting fecundity. Fecundity seems to be less dependent on body size than structural complexity. For example, the smaller (100–300 µm) nemertean produce 22 egg/batch, whereas the LF of placozoan (5 mm) is only one (Table 26.6). The structurally more complex brachiopods of ~ 40 mm size are more fecund (> 500–5,000 eggs) than the structurally simpler eutelic free-living nematodes (< 50 mm) laying < 300 eggs. When fecundity is plotted against the phylum, a positive linear relationship is apparent (Fig. 26.6). Arguably, it is the increase in structural complexity (rather than body size) with increasing cell number and tissue type that are required to support vitellogenesis and increasing fecundity. The simpler structural organization with less number of cells and tissue type limits fecundity per generation and results in reduced genetic diversity among the limited number of progenies. In its turn, the reduced genetic diversity limits species diversity in lower minor phyletics from Aorganomoprha to Hemocoelomata. The breaking point seems to commence with lophophores (e.g. Brachiopoda) and attains the peak in ascidians, irrespective of hermaphroditism in them. BF ranges from 500– 5,000 in iteroparous brachiopods and from 3,083 to 30,000 in iteroparous, solitary, non-clonal ascidians (Table 26.5). In colonials, sexual reproduction is supplemented with clonal multiplication. In sexual reproduction, egg production is almost nil in the least speciose entoprocts; it is at the modest level of 50 egg/zooid in *Botryllus schlosseri* and 2,012 egg-embryo/colony in *Ectenascidia turbinata* (Table 26.5) in the more speciose asicidians.

The lower minor phyletics, resource allocation is more for egg production than for spermatogenesis. As a result, sperm production is more limited than that of egg production. For example, the number of sperms produced is fewer

TABLE 26.6

Sexuality and fecundity in free-living minor phyletics. Par = Parthenogen, BF = Batch Fecundity, SF = Seasonal Fecundity, LF = Lifetime Fecundity, co = colony, mil = million

Particulars	Fecundity (egg no.)	Body Size
Placozoa, G: *Trichoplax adhaerens*, semelparous (Eitel et al., 2011), clonal multiplication = many	1	5 mm, 4–5 tissue types
Loricifera, G: *Rugiloricus carolinensis* (Kristensen, 2002) + parthenogenic cycle	4–12	80–800 µ
Cycliophora, G: *Symbium pandora* (Kristensen, 2002) + clonal multiplication	1 ?	~ 120 µ
Nemertea, G + a few ♀, *Prosorhochmus americanus* (Caplins and Turbeville, 2015); clonal multiplication in 4 *Lineus* species	22[BF]	100–300 µ, 60 m
Gnathostomulida, ♀: reciprocals	1–6	230 µ–3.6 mm
Rotifera, G: Par Monogonont fertilized 5.24 + unfertilized 22.4 Bdelloidea, *Adineta ricciae* (Ricci and Covino, 2005)	14–40 2–6 or 32	50 µ–2 mm
Kinorhyncha, G: *Echinoderes* (Huggins, 1974)	> 3	1 mm
Gastrotricha: Par, *Aspidophorus polystictos* (Balsamo and Todaro, 1988) 4 egg + ♀ 2.2 egg	6	0.1–1.5 mm
Nematoda, G: *Caenorhabditis elegans* (Klass, 1977); initially fertilized, subsequently sterile eggs (Schiemer, 1987)	273	< 50 mm
Priapulida, G (Wennberg, 2008)	Not known	10–40 cm
Sipuncula, G	Not known	2–60 cm
Echiura, G (Gibbs, 1975)	~ 80 oocytes	up to 1 m
Tardigrada, G: *Macrobiotus richtersi* (Altiero et al., 2006)	18–38; BF 10	1.2 mm
Onychophora, G: *Euperipetoides rowelli* (Sunnucks et al., 2000)	40	5 mm–15 cm
Entoprocta, G + a few ♀; Loxosomatidae (Mariscal, 1975) + clonal multiplication = 5	5[BF]/zooid	100 µ–5 mm
Phoronida, G + a few ♀; ♀ (Emig, 1982)	325[BF]; 633[BF]	6–200 mm
Bryozoa, ♀, *Flustrellidra hispida*, *Alcyonidium hirsutum* (Seed and Hughes, 1992) *Cupuladria exfragminis* (O'Dea, 2006)	52[BF], 210[LF]/co 50[BF], 245[LF]/co 1.5 times/ mo	up to 150 cm
Brachiopoda, free spawner oviparous Lophophore brooder Viviparous (see Fig. 24.2) (Meidlinger et al., 1998)	5,000[BF] 2,500[BF] 5,00[BF]	30–52 mm
Chaetognatha, ♀, *Paraspadella gotoi* (Goto and Yoshida, 1997)	227	40–100 mm

TABLE 26.6 Contd. ...

...TABLE 26.6 Contd.

Particulars	Fecundity (egg no.)	Body Size
Hemichordata (Pandian, 2018)	200–2,000	0.5 mm–2 m
Urochordata, ♀, Pelagic tunicates Benthic tunicates (see Table 26.5) Clonal tunicates	0.1–1.9 egg/d 3083 to 1 mil ~ 2100	600–1200 µ up to 6 cm

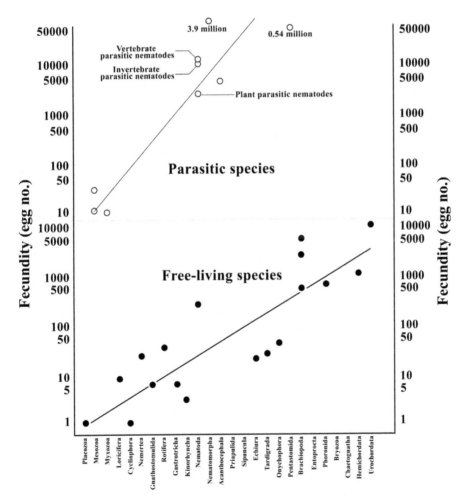

FIGURE 26.6

Fecundity in free-living (lower panel) and parasitic minor phyla (upper panel).

than the number of eggs in all the dicyemid species examined (see Furuya and Tsuneki, 2003). In mictic rotifers, 5 eggs alone are fertilized, in comparison to 22 eggs remaining unfertilized (Table 9.3). It is not known whether it is due to fewer sperm production in the mictic rotifers. In free-living nematodes, only a few sperm are produced so that the first few eggs are fertilized, but those arriving subsequently are not fertilized (Potts, 1910). Klass (1977) found that of 1,185 eggs produced by *Caenorhabditis elegans*, only 66% eggs are fertilized. In *C. elegans*, > 50% eggs remain unfertilized (Fig. 12.3A) and these sterile eggs do not develop. Due to protandry, the selfing *C. elegans* produce only 280 sperms and following irrevocable switching to the female phase, during which ~ 1,000 eggs are produced. As a result, there are only 0.3 sperms to fertilize an egg (Cutter, 2004). Perhaps, the sperm limitation is responsible for low fecundity in Mesozoa, Rotifera and free-living nematodes.

Parasitism

Tracing the transformation from symbiosis to parasitism in some turbellarians, Jennings (1977) defined that a parasite species may survive for a few days but not permanently in the absence of host(s). Remarkably, parasitism is manifested among minor phyla only in structurally relatively simpler Aorganomorpha, Pseudocoelomata and Hemocoelomata. The phyla Mesozoa (150 species), Myxozoa (2,200 species), Nematomorpha (360 species), Acanthocephala (1,100 species) and Pentastomida (144) include only parasitic species. Of 27,000 nematode species, 15, 13 and 31% are parasitic on plant (~ 4,100 species), invertebrates (~ 3,780 species) and vertebrate (8,366 species) host, respectively. On the whole, of 46,687 minor phyletic species, 20,200 or 43.3% are parasites. From these estimates, the following may be inferred: (1) Increase in structural complexity eliminates parasitism. (2) The parasites evoke structural degeneration in Aorganomorpha to a greater extent, and pseudocoelomate to a lesser extent; for example, structural degeneration is limited to minor changes in pharynx of nematodes. The structurally complex Pentastomida are not amenable to structural degeneration. Animal and fungal parasites tend to be mild and chronic, i.e. keep the host alive to draw nutrients as long as possible. The relatively simpler bacteria and virus are wild and epidemic to kill the host as early as possible.

The trend for the relation between fecundity and parasitic phyla is also positive and linear, but its slope level is so high that at a given level, the fecundity of parasitic minor phyletics is nearly 10–1,000 times more than that of free-living minor phyletics. The parasites are more fecund not only in the minor phyletics but also in the major phyletic. For example, BF of a free-living isopod *Idotea pelagica* is in the range of 17–37 eggs (see Johnson et al., 2001). But it increased to 250–750 eggs in the cymothoid parasitic isopod

TABLE 26.7

Sexuality and fecundity in parasitic minor phyletics. LF = Lifetime fecundity, BF = Batch fecundity

Phylum	Fecundity (egg no.)		Body Size
	Sexual	Clonal	
Aorganomorpha			
Mesozoa, G: *Ciliocinta sabellariae*; Kozloff (1971), Dicyemidae: ♀ *Corocyema*, (Furuya and Tsuneki, 2003) + clonal multiplication Orthonectida:	11–12 20		50 cells; 6–7 cell types
Myxozoa ♀ sexual cycle, 8 spores/plasmodium, no. of plasmodia produced by a species is not known + clonal multiplication	8	8 ?	
Psuedocoelomata			
Nematoda, Plant parasites	up to 1000^{LF}		
Mermithids	25,000^{LF}		5 mm–50 cm
Vertebrate parasites, oviparous	12,000^{LF}		25 cm–1 m
Vertebrate parasites, viviparous	10,000^{LF}		
Nematomorpha, Par, G: *Paragordius varius* (Hanelt, 2009)	3.5 million		15–25 cm to 2 m
Acanthocephala, G: *Corynosoma ceta* (Aznar et al., 2018)	4500		25 mm–65 cm
Pentastomida: G, *Porocephalus crotali* (Riley, 1981)	540,000/♀^{LF} 9275/♀/d^{BF}		80 mm–4 cm

Ichthyoxenus fushanensis with a direct life cycle and 100,000–250,000 eggs in the bopyrid parasite *Argeia pugattensis* involving an indirect life cycle with two intermediate hosts (Pandian, 2016). With assured food supply, the parasites allocate more of their resources for egg production. In this context, a point of interest is that the ratio between somatic and genital cell number is 85 : 15 (of 788 cells, only 74 are non-eutelic genital cells) in rotifers, whereas it is 65 : 35 (of 959 cells, as many as 340 are non-eutelic genital cells) for nematodes. Not surprisingly, the higher proportion of non-eutelic genital cells in nematodes has not only facilitated the free-living nematodes to be more fecund than the rotifers but has also led to more and more fecundity in plant (500–1,000 eggs), invertebrate (25,000 eggs) and vertebrate (10,000–12,000 eggs) parasitic nematodes (Table 26.7, Fig. 26.6). The presence of higher proportion of non-eutelic genital cells may also hold good for the other pseudocoelomate parasitic phyla Nematomorpha and Acanthocephala.

The following summarizes the new findings in minor phyletics. (1) Despite the fact that the minor phyla are aberrant clades terminating as blind offshoots, gonochorism commences with the simplest Placozoa and continues up to Hemocoelomata. With the breaking point commencing from lophophorates, it terminates in hermaphroditism in eucoelomates Chaetognatha and Ascidiacea. (2a) Regarding taxonomic distribution in minor phyla, hermaphroditism and parthenogenesis mutually eliminate each other. (2b) Similarly, parthenogenesis and clonals also mutually eliminate each other. (2c) However, clonal multiplication occurs both in gonochores and hermaphrodites. (3) The minor phyletics are less speciose (1,795 species/phylum) than that (157,066 species/phylum) of major phyletics. The reasons for the reduced species diversity are traced to (i) the incidence of 6.4% parthenogenesis plus 21.6% clonals; (ii) structural complexity also eliminates the manifestation of parthenogens in higher minor phyletics and (iii) the fecundity is low in minor phyletics (from Placozoa to Hemocoelomata), especially due to eutelism in Mesozoa and Pseudocoelomata; in them, sperm production is less than that of egg production; as a result, a large number of their eggs remain unfertilized. (4) With < 100 selfing species, hermaphroditism may not be a reason for species-poverty in minor phyla. Colonial (or fragment) fusion, and colonial degeneration and rejuvenization by regeneration may have enriched the species number in asicidians and possibly in bryozoans. (5) Structural simplicity facilitates the manifestation of parasitism in lower minor phyla. On the other hand, structural complexity eliminates it in higher minor phyla. (6) A higher proportion of non-eutelic genital cells in nematodes, and possibly in other pseudocolomate parasitic phyla facilitates more fecundity. The smaller number of egg and progeny production per generation has reduced genetic diversity and thereby species diversity in minor phyla, especially in free-living clades.

27

References

Abebe, E., Decraemer, W. and De Ley, P. 2008. Global diversity of nematodes (Nematoda) in freshwater. Hydrobiologia, 595: 67–68.

Abeloos, M. 1951. Morphongenese des colonies du *Bugula*. C R Acad Sci, Paris, 232: 652–656.

Adiyodi, K.G. and Adiyodi, R. 1983. *Reproductive Biology of Invertebrates: Oogenesis, Oviposition and Oosorption*. Wiley, UK. Vol 1, p 796.

Adiyodi, K.G. and Adiyodi, R. 1984. *Reproductive Biology of Invertebrates: Spermatogenesis and Sperm Function*. Oxford & IBH Publishers, New Delhi, Vol 2, p 718.

Adiyodi, K.G. and Adiyodi, R. 1988. *Reproductive Biology of Invertebrates: Accessory Sex Glands.* Oxford & IBH Publishers, New Delhi, Vol 3, p 542.

Adiyodi, K.G. and Adiyodi, R. 1989. *Reproductive Biology of Invertebrates: Fertilization, Development and Parental Care: Porifera–Annelida*. Oxford & IBH Publishers, New Delhi, Vol 4A, p 463.

Adiyodi, K.G. and Adiyodi, R. 1990. *Reproductive Biology of Invertebrates: Fertilization, Development and Parental Care: Clitellata–Urochordata*. Oxford & IBH Publishers, New Delhi, Vol 4B, p 527.

Adiyodi, K.G. and Adiyodi, R. 1993. *Reproductive Biology of Invertebrates: Sexual Differentiation and Behaviour*. Oxford & IBH Publishers, New Delhi, Vol 5, p 536.

Adiyodi, K.G. and Adiyodi, R. 1994. *Reproductive Biology of Invertebrates: Asexual Propagation and Reproductive Strategies–Porifera–Mollusca*. Oxford & IBH Publishers, New Delhi, Vol 6A, p 410.

Adiyodi, K.G. and Adiyodi, R. 1995. *Reproductive Biology of Invertebrates: Asexual Propagation and Reproductive Strategies–Clitellata–Cephalochordata*. Oxford & IBH Publishers, New Delhi, Vol 6B, p 432.

Adiyodi, K.G. and Adiyodi, R. 2000. *Reproductive Biology of Invertebrates: Progress in Male Gamete Ultrastructure and Phylogeny*. Oxford & IBH Publishers, New Delhi, Vol 9B, p 450.

Adiyodi, K.G. and Adiyodi, R. 2001. *Reproductive Biology of Invertebrates: Progress in Developmental Endocrinology*. Oxford & IBH Publishers, New Delhi, Vol 10A, p 332.

Adoutte, A., Balavoine, G., Lartillot, N. et al. 2000. The new animal phylogeny: Reliability and implications. Proc Natl Acad Sci, USA, 97: 4453–4456.

Adrianov, A.V. and Malakhov, V.V. 1996. *Priapulida: Morphology, Development, Phylogeny and Classification*. KMK Scientific Press, Moscow, p 268.

Adrianov, A.V. and Malakhov, V.V. 1999. *Cephalorhyncha of the World Ocean*. KMK Scientific Press, Moscow, p 328.

Adrianov, A.V. and Maiorova, A.S. 2010. Reproduction and development of common species of peanut worms (Sipuncula) from the Sea of Japan. Russ J Mar Biol, 36: 1–15.

Adrianov, A.V., Maiorova, A.S. and Malakhov, V.V. 2011. Embryonic and larval development of the peanut worm *Phascolosoma agassizii* (Keferstein 1867) from the Sea of Japan (Sipuncula: Phascolosomatidea). Invert Reprod Dev, 55: 22–29.

Ahmad, I. and Kaur, H. 2018. Prevalence, site and tissue preference of myxozoan parasites infecting gills of cultured fingerlings of Indian major carps in District Fatehgarh Sahib, Punjab (India). J Parasitol Dis, doi.org/10.1007/s12639-018-1035-6.

Alberti, G. and Storch, V. 1988. Internal fertilization in a meiobenthic priapulid worm: *Tubiluchus philippinensis* (Tubiluchidae, Priapulida). Protoplasma, 143: 193–196.

Ali, J.H., Riley, J. and Self, J.T. 1981. A revision of the taxonomy of the blunt-hooked *Raillietiella*, pentastomid parasites of African, South-East-Asian and Indonesian lizards, with a description of a new species. Syst Parasitol, 3: 193–207.

Altiero, T., Rebecchi, L. and Bertolani, R. 2006. Phenotypic variations in the life history of two clones of *Macrobiotus richtersi* (Eutardigrada, Macrobiotidae). Hydrobiologia, 558: 33–40.

Alvarez-Presas, M. and Riutort, M. 2014. Planarian (Platyhelminthes, Tricladida) diversity and molecular markers: A new view of an old group. Diversity, 6: 323–338.

Alvarino, A. 1990. Chaetognatha. In: *Reproductive Biology of Invertebrates*. (eds) Adiyodi, K.G. and Adiyodi, R.G., Oxford and IBH Publishing, New Delhi, Vol IV, Part B, pp 255–282.

Ament-Velasquez, S.L., Figuet, E., Ballenghien, M. et al. 2016. Population genomics of sexual and asexual lineages in fissiparous ribbon worms (*Lineus*, Nemertea): hybridization, polyploidy and the Meselson effect. Mol Ecol, 25: 3356–3369.

Anderson, R.C. 2000. *Nematode Parasities of Vertebrates*. CABI Publlishing, Wallingford, UK, p 672.

Anderson, R.V. and Darling, H.M. 1964. Embryology and reproduction of *Ditylenchus destructor* Thorne with emphasis on gonad development. Proc Helminthol Soc Wash, 31: 240–256.

Andrade, S., Strand, M., Schwartz, M. et al. 2012. Disentangling ribbon worm relationships: multi-locus analysis supports traditional classification of the phylum Nemertea. Cladistics, 28: 141–159.

Andrade, S.C., Montenegro, H., Strand, M. et al. 2014. A transcriptomic approach to ribbon worm systematic (Nemertea): resolving the Pilidiophora problem. Mol Biol Evol, 31: 3206–3215.

Annandale, N. 1911. Fauna of British India. In: *Freshwater Sponges, Hydroids and Polyzoa*. Good Press, p 274.

Antipov, A.A., Bakhur, T.I., Feshchenko, D.V. et al. 2018. Earthworms (Lumbricidae) as intermediate hosts of lung nematodes (Metastrongylidae) of swine in Kyiv and Zhytomyr regions of Ukraine. Vestnik Zoologii, 52: 59–64.

Atkins, D. 1932. The Loxostomatidae of the Plymouth area including *L. obesum* sp Nov. Q J Microsc Soc, 75: 321–391.

Awachie, J.B.E. 1966. The development and life history of *Echinorhynchus truttae* Schrank 1788 (Acanthocephala). J Helminthol, 40: 11–32.

Awati, P.R. and Pradhan, L.B. 1936. The anatomy of *Dendrostoma signifier* Selenka and de Man. J Univ Bombay, 4: 114–131.

Ayeh-Kumi, P.F., Quarcoo, S., Kwakye-Nuako G. and Kretchy, J.-P. 2009. Prevalence of intestinal parasitic infections among food vendors in Accra, Ghana. J Trop Med Parasitol, 32: 1–8.

Aymes, Y. 1956. Criossance phototropique chez les Bryozoaires du genre *Bugula*. C R Acad Sci, Paris, 242: 1237–1238.

Aznar, F.J., Hernandez-Orts, J.S., Velez-Rubio, G. et al. 2018. Reproductive inequalities in the acanthocephalan *Corynosoma cetaceum*: looking beyond 'crowding' effects. Parasites & Vectors, 11: 1–10.

Baird, M.J., Lee, D.E. and Lamare, M.D. 2013. Reproduction and growth of the terebratulid brachiopod *Liothyrella neozelanica* Thomson, 1918 from Doubtful Sound, New Zealand. Biol Bull, 225: 125–136.

Bak, R.P.M., Sybesma, J. and van Duyl, F.C. 1981. The ecology of tropical compound ascidian *Trididemnum solidum*. II. Abundance, growth and survival. Mar Evol Prog Ser, 6: 43–52.

Balsamo, M. and Todaro, M.A.D. 1988. Life history traits of two chaetonotids (Gastrotricha) under different experimental conditions. Invert Reprod Dev, 14: 161–176.

Balsamo, M., d'Hondt, J-L., Kisielewski, J. et al. 2015. Fauna Europaea: Gastrotricha. Biodiv Data J, 3: e5800, DOI: 10.3897/BDJ.3.e5800.

Banaja, A.A., James, J.L and Riley, J. 1976. Some observations on egg production and autoreinfection of *Reighardia sternae* (Diesing, 1894), a pentastomid parasite of the herring gull. Parasitology, 72: 81–91.

Banerjee, P., Basu, S. and Modak, B.K. 2016. Prevalence of myxozoan parasites in fishes from River Damodar at Mejia area, Bankura, West Bengal. J Environ Sociobiol, 13: 131–135.

Bauer, R.T. 2005. Cost of maleness on brood production in the shrimp *Lysmata wurdemanni* (Decapoda: Caridea: Hippolytidae) a protandric simultaneous hermaphrodite. J Mar Biol Ass UK, 85: 101–106.

Beauchamp, K.A., Gay, M., Kelley, G.O. et al. 2002. Prevalence and susceptibility of infection to *Myxobolus cerebralis*, and genetic differences among populations of *Tubifex tubifex*. Dis Aquat Org, 51: 113–121.

Bell, D.A. and Beverley-Burton, M. 1980. Prevalence and intensity of *Capillaria catostomi* (Nematoda: Trichuroidea) in white sucker (*Catostomus commersoni*) in southern Lake Huron, Canada. Env Biol Fish, 5: 267–271.

Bell, G. 1982. *The Masterpiece of Nature: The Evolution and Genetics of Sexuality*. University of California Press, Berkeley, p 634.

Benesh, D.P. and Valtonen, E.T. 2007. Sexual differences in larval life history traits of acanthocephalan cystacanths. Int J Parasitol, 37: 191–198.

Berec, L., Schembri, P.J. and Boukal, D.S. 2005. Sex determination in *Bonellia viridis* (Echiura: Bonelliidae): population dynamics and evolution. Oikos, 108: 473–484.

Bernard, G.C., Egnin, M. and Bonsi, C. 2017. The impact of plant-parasitic nematodes on agriculture and methods of control. In: *Nematology: Concepts, Diagnosis and Control*. (eds) Shah, M.M. and Mohamood, M., Books on Demand, pp 121–151.

Berrill, N.J. 1935. Studies in tunicate development. Part IV—Asexual Reproduction. Phil Trans R Soc London, 225: 327–379.

Bertolani, R. 2001. Evolution of the reproductive mechanisms in tardigrades—A review. Zool Anz, 240: 247–252.

Bertolani, R., Guidetti, R., Jonsson, K.I. et al. 2004. Experiences on dormancy in tardigrades. J Limnol, 63: 16–25.

Bierne, J. and Rue, G. 1979. Endocrine control of reproduction in two rhynchocoelan worms. Int J Invert Reprod, 1: 109–120.

Bierne, J. 1990. *Lineus* as a model for studying developmental processes in animals reconstructed from adult pieces. Int J Dev Biol, 34: 245–253.

Bird, A.F. and Bird, J. 1991. *The Structure of Nematodes*. Academic Press, San Diego, California, p 335.

Birky, C.W. and Gilbert, J.J. 1971. Parthenogenesis in rotifers: The control of sexual and asexual reproduction. Am Zool, 11: 245–266.

Birky, C.W. 2010. Positively negative evidence for asexuality. J Hered, 101S: 42–45.

Bishop, D.D. and Sommerfeldt, A.D. 1999. Not like *Botryllus*: indiscriminate post-metamorphic fusion in a compound ascidian. Proc R Soc Lond, 266B: 241–248.

Blackstone, N.W. and Jasker, B.D. 2003. Phylogenetic considerations of clonality, colonality, and mode of germline development in animals. J Exp Zool, 297B: 35–47.

Blackwelder, R.E. and Garoian, G.S. 1986. *Handbook of Animal Diversity*. CRC Press, USA, p 555.

Blaxter, M. and Koutsovoulos, G. 2015. The evolution of parasitism in Nematoda. Parasitology, 142S: 26–39.

Boag, B. and Yeates, G.W. 1998. Soil nematode biodiversity in terrestrial ecosystems. Biodiv Conserv, 7: 617–630.

Bock, P.E. and Gordon, D.P. 2013. Phylum Bryozoa Ehrenberg, 1831. Zootaxa, 3703: 67–74.

Bone, E.K. and Keough, M.J. 2005. Responses to damage in an arborescent bryozoan: effects of injury location. J Exp Mar Biol Ecol, 324: 127–140.

Borg, F. 1947. Zur Kenntnis der Okologie und Lebenszyklus von *Elektra crustalents*. Zool Bid, Uppsala, 25: 344–377.

Borisanova, A. and Potanina, D. 2016. A new species of *Coriella*, *Coriella chernyshevi* n sp (Entoprocta, Barentsiidae), comments on the genera *Coriella* and *Pedicellinopsis*. Zootaxa, 4184: 376, DOI: 10.11646/zootaxa.4184.2.9.

Bourlat, S., Nielsen, C., Economou, A.D. and Telford, M.J. 2008. Testing the new animal phylogeny: A phylum level molecular analysis of the animal kingdom. Mol Phylogen Evol, 49: 23–31.

Boyle, M.J. and Seaver, E.C. 2010. Expression of FoxA and GATA transcription factors correlates with regionalized gut development in two lophotrochozoan marine worms: *Chaetopterus* (Annelida) and *Themiste lageniformis* (Sipuncula). Evol Dev, 1: 2, 10.1186/2041-9139-1-2.

Boyle, M.J. and Rice, M.E. 2014. Sipuncula: an emerging model of spiralian development and evolution. Int J Dev Biol, 58: 485–499.

Braem, F. 1890. Untersuchungen uber die Bryozoan des sussen wassers. Zoologica, Stuttgart, Vol 2, Part 6.

Braem, F. 1951. Uber *Victorella* und einige ihre nachsten Verwandten. Zoologica Stuttgart, 102: 1–59.

Brianti, E., Gaglio, G., Napoli, E. et al. 2012. New insights into the ecology and biology of *Acanthocheilonema reconditum* (Grassi, 1889) causing canine subcutaneous filariosis. Parasitology, 139: 330–336.

Brien, P. 1936. Reproduction asexuee des phylactolaemates. Mem Mus Hist Natur Belgique, 3: 569–625.

Brien, P. 1953. Etude sur les Phylactolemates. Annales Soc R Zool Belgique, 84: 301–444.

Bronstein, G. 1938. Mecanisme de la formation du polypide chez *Membranipora*. C R Acad Sci, Paris, 207: 506–508.

Brook, H.J., Rawlings, T.A. and Davies, R.W. 1994. Protogynous sex change in the intertidal isopod *Gnorimosphaeroma oregonense* (Crustacea: Isopoda). Biol Bull, 187: 99–111.

Brown, C.J.D. 1933. A limnological study of certain freshwater Polyzoa with special reference to their statoblasts. Trans Am Microsc Soc, 52: 271–316.

Brown, F.D. and Swalla, B.J. 2012. Evolution and development of budding by stem cells: ascidian colonality as a case study. Dev Biol, 369: 151–162.

Buchner, H., Mutschler, C. and Kieleble, H. 1967. Die Determination der Manchen und Dauereiproduktion bei *Asplanchna sieboldi*. Biol Zentrabl, 86: 599–621.

Budd, G.E. 2001. Tardigrades as 'stem-group arthropods': the evidence from the Cambrian fauna. Zool Anz, 240: 265–279.

Bulow, C. 1883. Uber freiwillige und kunstilche Teilung mit nachfolgender Regeneration bei Coelenteraten, Echinoderman und Wurmenn. Biol Zentralbl, 3: 14–20.

Bultzer, F. 1931. Entwicklungsmechanische Untersuchungen an *Bonellia viridis*. Die Abhängigkeit der Entwicklungsgeschwindigkeit und des Entwicklungsgrades der männlichen Larvae von der Dauer der Russel-Parasitismus. Rev Suisse Zool, 38: 361–371.

Burlingame, P.L. and Chandler, A.C. 1941. Host parasite relations of *Moniliformis dubius* (Acanthocephala) in albino rats, and the environmental nature of resistance to single and superimposed infections with this parasite. Am J Hyg, 33: 1–21.

Butterworth, P.E. and Barrett, J. 1985. Anaerobic metabolism in the free-living nematode *Panagrellus redivivus*. Physiol Zool, 58: 9–17.

Cable, R.M. 1971. Parthenogenesis in parasitic helminths. Am Zool, 11: 267–272.

Cairncross, S., Muller, R. and Zagaria, N. 2002. Dracunculiasis (Guinea worm disease) and the eradication initiative. Clinc Microbiol Rev, 15: 223–246.

Callaghan, T.P. and Karlson, R.H. 2002. Summer dormancy as a refuge from mortality in the freshwater bryozoan *Plumatella emarginata*. Oecologia, 132: 51–59.

Canning, E.U., Curry, A., Feist, S. and Longshaw, M. 2000. A new class and order of myxozoans to accommodate parasites of bryozoans with ultrastructural observations on *Tetracapsula bryosalmonae* (PKX organism). J Eukary Microbiol, 47: 456–468.

Canning, E.U. and Okamura, B. 2004. Biodiversity and evolution of the Myxozoa. Adv Parasitol, 56: 10.1016/S0065-308X(03)56002-X.

Caplins, S.A. and Turbeville, J.M. 2015. High rates of self-fertilization in a marine ribbon worm (Nemertea). Biol Bull, 229: 255–264.

Cappucci, D.T. 1976. The biology of *Gordius robustus* Leidy with a host list and summary of the public health importance of the Gordioidea. Ph.D. Dissertation. University of California, San Francisco. San Francisco, California, p 240.

Carlisle, D.B. 1961. Locomotory powers of adult ascidians. Proc Zool Soc Lond, 136: 141–146.

Carre, D., Djediat, C. and Sardet, C. 2002. Formation of a large Vasa-positive granule and its inheritance by germ cells in the enigmatic chaetognaths. Development, 129: 661–670.

Caullery, M. and Lavallee, A. 1908a. La fecondation et le development de l'oeuf chez un Orthonectide (*Rhopalura ophiocomae*). CR Acad Sci, Paris, 146: 40–43.

Caullery, M. and Lavallee, A. 1908b. La fecundation et le development de l'oeuf des Orthoenectides. I. *Rhopalura ophiocomae*. Arch Zool Exp Gen, 4: 421–469.

Caullery, M. and Lavallee, A. 1910. Recherches experimentale sur les phases initiales de 1 'infection d' une Ophiure (*A. squamata*) par un Orthonectide (*R. ophiocomae*). CR Acad Sci, Paris, 150: 1781–1783.

Chapman, A.D. 2009. *Number of Living Species in Australia and the World*. Department of the Environment, Water, Heritage and the Arts. Government of Australia, p 80.

Charlesworth, D. and Willis, J.H. 2009. The genetics of inbreeding depression. Nat Rev Genet, 10: 783–796.

Charwat, S.M., Davies, K.A. and Hunt, C.H. 2000. Impact of rhabditid nematode on survival and fecundity of *Cernuella virgata* (Mollusca: Helicidae). Biocontrol Sci Tech, 10: 147–155.

Chavhan, P.B., Khan, L.A., Raut, P.A. et al. 2008. Prevalence of nematode parasites of ruminants at Nagpur. Vet World, 1: 140.

Chia, F.S. and Warwick, R.M. 1969. Assimilation of labeled glucose from seawater by marine nematodes. Nature, 224: 720–721.

Chintala, M.M. and Kennedy, V.S. 1993. Reproduction of *Stylochus ellipticus* (Platyhelminthes: Polycladida) in response to temperature, food, and presence or absence of a partner. Biol Bull, 185: 373–387.

Chitwood, D.J. 2003. Research on plant-parasitic nematode biology conducted by the United States Department of Agriculture-Agricultural Research Service. Pest Manag Sci, 59: 748–753.

Christoffersen, M.L. and de Assis, J.E. 2015. Pentastomida. Revista IDE, 98b: 1–10.

Chuang, S.H. 1959. The breeding season of the brachiopod, *Lingula unguis* (L.). Biol Bull, 117: 202–207.

Chuang, S.H. 1990. Brachipoda. In: *Reproductive Biology of Invertebrates: Fertilization, Development and Parental Care*. (eds) Adiyodi, K.G. and Adiyodi, R.G., Oxford & IBH Publishers, New Delhi, Vol 4B, pp 211–254.

Clark, M.S., Denekamp, N.Y., Thorne, M.A.S. et al. 2012. Long-term survival of hydrated resting eggs from *Brachionus plicatilis*. PLoS ONE, 7: e29365.

Cloney, R.A. 1990. Larval tunic and the function of the test cells in ascidians. Acta Zool, 71: 151–159.

Coe, W.R. 1943. *Biology of the Nemerteans of the Atlantic Coast of North America*. Trans Connecticut Acad Arts Sci, The University of California, p 327.

Cribb, T.H., Bray, R.A. and Littlewood, D.T.J. 2001. The nature and evolution of the association among digeneans, molluscs and fishes. Int J Parasitol, 31: 997–1011.

Crompton, D.W.T. 1976. Entry into the host and site selection. In: *Ecological Aspects of Parasitology*. (ed) Kennedy, C.R., North Holland Publisher, Amsterdam, pp 41–73.

Crompton, D.W.T., Singhvi, A. and Keymer, A. 1982. Effects of host dietary fructose on experimentally stunted *Moniliformis* (Acanthocephala). Int J Parasitol, 12: 117–121.

Crompton, D.W.T. 1985. Reproduction. In: *Biology of the Acanthocephala*. (eds) Crompton, D.W.T. and Nickol, B.B., Cambridge University Press, pp 213–271.

Crompton, D.W.T., Keymer, A.E., Walters, D.E. et al. 1988. Factors influencing the fecundity of *Moniliformis moniliformis* (Acanthocephala): constant dose and varied diet. J Zool Lond, 214: 221–234.

Cutler, E.B. 1994. *The Sipuncula: Their Systematics, Biology and Evolution*. Cornell University Press, p 453.

Cutter, A.D. 2004. Sperm-limited fecundity in nematodes: How many sperm are enough? Evolution, 58: 651–655.

Das, N.K. 1968. Developmental features and synthetic patterns of male germ cells of *Urechis caupo*. Wilhelm Roux Arch Entwickl Mech Org, 161: 325–335.

Davis, F.C. 1989. Echiura. In: *Reproductive Biology of Invertebrates*. (eds) Adiyodi, K.G. and Adiyodi, R.G., Oxford & IBH Publishers, New Delhi, Vol 4A, pp 349–381.

Dawydoff, C. 1924. Sur le retour d'une nemerte en inianiten a un etat embryonaire. C R Acad Sci, Paris, 179: 361–363.

Dawydoof, C. 1959. Classe des Echiuriens. In: *Traitie de Zoologie*. (ed) Grasse, P.P., Masson et cie, Paris, Vol 6, pp 855–907.

Debortoli, N., Li, X., Eyres, I. et al. 2016. Genetic exchange among bdelloid rotifers is more likely due to horizontal gene transfer than to meiotic sex. Curr Biol, 26: 1–10.

Denver, D., Clark, K.A. and Raboin, M.J. 2011. Reproductive mode evolution in nematode: insight from molecular phylogenies and recently discovered species. Mol Phylogenet Evol, 61: 584–592.

Dobson, A.P. 1986. Inequalities in the individual reproductive success of parasites. Parasitology, 92: 675–682.

Dong, L., Sanad, M., Wang, Yi. et al. 2014. Mating clusters in the mosquito parasitic nematode, *Strelkovimermis spiculatus*. J Invert Pathol, 117: 19–25.

Durante, K.M. and Sebens, K.P. 1994. Reproductive ecology of the ascidians *Molgula citrine* Alder & Hancock, 1848 and *Aplidium glabrum* (Verrill, 1871) from the Gulf of Maine, USA. Ophelia, 39: 1–21.

Dybing, N.A., Fleming, P.A. and Adams, P.A. 2013. Environmental conditions predict helminth prevalence in red foxes in Western Australia. Int J Parasitol: Parasites Wildlife, 2: 165–172.

Eitel, M., Guidi, L., Hadryus, H. et al. 2011. New insights into placozoan sexual reproduction and development. PLoS ONE, 6: e19639.

Eitel, M., Osigus, H-J., DeSalle, R. and Schierwater, B. 2013. Global diversity of the Placozoa. PLoS ONE, 8: e57131.

El-Sayed, A.A.M., El-Damhogy, Kh.A., Hanafy, M.H. and El-Kareem, A.F.G. 2018. Reproductive biology of the solitary ascidian, *Herdmania momus* (Ascidiacea: Hemichordata) from Hurghada Coasts, Red Sea, Egypt. Egypt Acad J Biol Sci, 10: 47–60.

Emig, C.C. 1982. The biology of Phoronida. Adv Mar Biol, 19: 1–89.

Emig, C.C. 1990. Phoronida. In: *Reproductive Biology of Invertebrates*. (eds) Adiyodi, K.G. and Adiyodi, R.G., Oxford & IBH Publishers, New Delhi, Vol 4, Part B, pp 165–184.

Emschermann, P. 1971. *Loxomespilon perezi*–ein Entoproctenfund im Mittelatlantik. Uberlegungen zur Benthosbesiedlung der Grossen Meteorbank. Mar Biol, 9: 51–62.

Eriksson, B.J. and Tailt, N.N. 2012. Early development in the velvet worm *Euperipatoides kanangrensis* Reid 1996 (Onychophora: Peripatopsidae). Arthro Struc Dev, 41: 483–493.

Eszterbauer, E., Atkinson, S., Diamant, A. et al. 2015. Myxozoan life cycles: Practical approaches and insights. In: *Myxozoan Evolution, Ecology and Development*. Springer International, Switzerland, p 175–198.

Eyres, I., Boschetti, C., Crisp, A. et al. 2015. Horizontal gene transfer in bdelloid rotifers is ancient, ongoing and more frequent in species from desiccating habitats. BMC Biol, 13. Doi: 10.1186/s12915-015-0202-9.

Feist, S.W. and Longshaw, M. 2006. Phylum Myxozoa. In: *Fish Diseases and Disorders*. (ed) Woo, P.T.K., CABI International, London. Vol 1, pp 230–296.

Feist, S.W., Morris, D.J., Alama-Bermejo, G. and Holzer, A.S. 2015. Cellular processes in Myxozoans. In: *Myxozoan Evolution, Ecology and Development*. Springer International, Switzerland, p 139–154.

Fell, P.E. 1974. Porifera. In: *Reproduction of Marine Invertebrates*. (eds) Giese, A.C. and Pearse, J.S., Academic Press, New York, Vol 1, pp 51–132.

Ferris, J.C. 1932. A comparison of the life histories of mictic and amictic females in the rotifer, *Hydatina senta*. Biol Bull, 63: 443–455.

Figuerola, B. and Avila, C. 2019. The phylum Bryozoa as a promising source of anticancer drugs. Mar Drugs, 17: 477. doi:10.3390/md17080477.

Flot, J.-F., Hespeels, B., Li, X. et al. 2013. Genomic evidence for ameiotic evolution in the bdelloid rotifer *Adineta vaga*. Nature, doi: 10.1038/nature12326.

Fontaneto, D., De Smet, W.H. and Melone, G. 2008. Identification key to the genera of marine rotifers worldwide. Meiofauna Marine, Munchen, 16: 1–200.

Fourie, H., Mc Donald, A.H., Steenkamp, S. and De Waele, D. 2017. Nematode pests of leguminous and oilseed crops. In: *Nematology in South Africa: A View from the 21st Century*. (eds) Fourie, H. et al. Springer, Switzerland. pp 201–230.

Freeman, G. 1995. Regional specification during embryogenesis in the inarticulate brachiopod *Glottidia*. Dev Biol, 172: 15–36.

Freeman, G. 1999. Regional specification during embryogenesis in the inarticulate brachiopod *Discinisca*. Dev Biol, 209: 321–339.

Freeman, G. 2000. Regional specification during embryogenesis in the craniiform brachiopod *Crania anomala*. Dev Biol, 227: 219–238.

Fujiwara, R.T., Zhan, B., Mendez, S. et al. 2007. Reduction of worm fecundity and canine host blood loss mediates protection against hookworm infection elicited by vaccination and recombinant. Clinic Vac Immunol, 14: 281–287.

Funch, P. and Kristensen, R.M. 1995. Cycliophora is a new phylum with affinities to Entoprocta and Ectoprocta. Nature, 378: 711–714.

Furuya, H., Tsuneki, K. and Koshida, Y. 1992. Two new species of the genus *Dicyema* (Mesozoa) from octopuses of Japan with notes on *D. misakiense* and *D. acuticephalum*. Zool Sci, 9: 423–437.

Furuya, H., Tsuneki, K. and Koshida, Y. 1996. The cell lineages of two types of embryo and a hermaphroditic gonad in dicyemid mesozoans. Dev Growth Differ, 38: 453–463.

Furuya, H., Hochberg, F.G. and Tsuneki, K. 2001. Developmental patterns and cell lineages of vermiform embryos in dicyemid mesozoans. Biol Bull, 201: 405–416.

Furuya, H. and Tsuneki, K. 2003. Biology of dicyemid mesozoans. Zool Sci, 20: 519–532.

Garcia-Cagida, A., Hernandez-Zanuy, A. and Cardenas, A. 2005. Fecundidad y primeras fases del desarrollo larval de la ascidia *Ecteinascidia turbinata* (Ascidiacea: Perophoridae) en Cuba. Bol Invest Mar Cost, 34: 141–159.

Ghiselin, M.T. 1969. The evolution of hermaphroditism among animals. Q Rev Biol, 44: 189–208.

Gibbs, P.E. 1975. Gametogenesis and spawning in a hermaphroditic population of *Golfingia minuta* (Sipuncula). J Mar Biol Ass UK, 55: 69–82.

Giese, A.C. and Pearse, J.S. 1974. *Reproduction of Marine Invertebrates: Acoelomata and Pseudocoelomata*. Academic Press, New York, Vol 1, p 546.

Giese, A.C. and Pearse, J.S. 1975a. *Reproduction of Marine Invertebrates: Entoprocts and Lesser Coelomates*. Academic Press, New York, Vol 2, p 352.

Giese, A.C. and Pearse, J.S. 1975b. *Reproduction of Marine Invertebrates: Annelida and Echiura*. Academic Press, New York, Vol 3, p 347.

Giese, A.C. and Pearse, J.S. 1991. *Reproduction of Marine Invertebrates: Echinoderms and Locophorata*. Boxwood Press, Vol 6, p 808.

Gilbert, J.J. 1971. Some notes on the control of sexuality in the rotifer *Asplanchna sieboldi*. Limnol Oceanogr, 16: 309–319.

Gilbert, J.J. 1974. Dormancy in rotifers. Trans Amer Microsc Soc, 93: 490–513.

Gilbert, J.J., Birky, C.W. Jr. and Wurdak, E.S. 1979. Taxonomic relationships of *Asplanchna brightwelli*, *A. intermedia*, and *A. sieboldi*. Arch Hydrobiol, 87: 224–242.

Gilbert, J.J. 1993. Rotifera. In: *Reproductive Biology of Invertebrates*. (eds) Adiyodi, K.G. and Adiyodi, R.G., Oxford & IBH Publishers, New Delhi, Vol 6, Part A, pp 231–263.

Gilbert, J.J. 1995. Structure, development and induction of a new diapause stage in rotifers. Freshwat Biol, 34: 263–270.

Gilbert, J.J. 2002. Endogenous regulation of environmentally induced sexuality in a rotifer: a multigenerational parental effect induced by fertilization. Freshwat Biol, 47: 1633–1641.

Gilbert, J.J. and Schroder, T. 2004. Rotifers from diapausing fertilized eggs: Unique features and emergence. Limnol Oceanogr, 49: 1341–1354.

Gilbert, J.J. 2010. Effect of food concentration on the production and viability of resting eggs of the rotifer *Brachionus*: implications for the timing of sexual reproduction. Freshwat Biol, 55: 2437–2446.

Gilbert, J.J. and Dieguez, M.C. 2010. Low crowding threshold for induction of sexual reproduction and diapause in a Patagonian rotifer. Freshwat Biol, 55: 1705–1718.

Gladyshev, E. and Meselsen, M. 2008. Extreme resistence of bdelloid rotifers to ionizing reduction. Proc Natl Acad Sci USA, 105: 5139–5144.

Glazier, D.S. 1992. Effects of food, genotype, and maternal size and age on offspring investment in *Daphnia magna*. Ecology, 79: 1371–1381.

Glime, J.M. 2017a. Invertebrate rotifers. In: *Bryophyte Ecology: Biological Interaction*. Michigan Tech University and Int Ass Bryol, 4.5: 1–19.

Glime, J.M. 2017b. Tardigrade survival. In: *Bryophyte Ecology: Biological Interaction*. Michigan Tech University and Int Ass Bryol, 5.1: 1–21.

Glime, J.M. 2017c. Invertebrates: Nematodes. In: *Bryophyte Ecology: Biological Interaction*. Michigan Tech University and Int Ass Bryol, 4.3: 1–4.

Goransson, U., Jacobsson, E., Strand, M. and Andersson, H.S. 2019. The toxins of nemertean worms. Toxins, 11: 120. DOI: 10.3390/toxins11020120.

Gordon, D.P., Clark, AG. and Harper, J.F. 1987. Bryozoa. In: *Animal Energetics*. (eds) Pandian, T.J. and Vernberg, F.J., Academic Press, New York, pp 173–199.

Goto, T. and Yoshida, M. 1997. Growth and reproduction of the benthic arrowworm *Paraspadella gotoi* (Chaetognatha) in laboratory culture. Invert Reprod Dev, 32: 201–207.

Gould, M.C. and Stephano, J.L. 1996. Fertilization and parthenogenesis in *Urechis caupo* (Echiura). Invert Reprod Dev, 30: 1–3.

Gould-Somero, M. 1975. Echiura. In: *Reproduction of Marine Invertebrates*. (eds) Giese, A.C. and Pears, J.S., Academic Press, New York, Vol 3, pp 277–312.

Grell, K.G. 1972. Eibildung und Furchtung von *Trichoplax adhaerens* F.E. Schultz (Placozoa). Z Morphol Tiere, 73: 297–314.

Grigor, J.J., Schmid, M.S. and Fortier, L. 2017. Growth and reproduction of the chaetognaths *Eukrohnia hamata* and *Parasagitta elegans* in the Canadian Arctic Ocean: capital breeding versus income breeding. J Plankton Res, 39: 910–929.

Gross, V., Treffkorn, S. and Mayer, G. 2015. Tardigrada. In: *Evolutionary Developmental Biology of Invertebrates*: *Ecdysozoa I: Non-Tetraconata*. (ed) Wanninger, A., Springer Verlag, Wein. Vol 3, pp 35–52.

Gruhl, A. 2015. Myxozoa. In: *Evolutionary Developmental Biology of Invertebrates*. (ed) Wanninger, A., Springer Verlag, Wein. Vol 1, pp 165–178.

Gruhl, A. and Okamura, B. 2015. Tissue characteristics and development in myxozoa. In: *Myxozoan Evolution, Ecology and Development*. (ed) Wanninger, A., Springer International, Switzerland, p 155–174.

Guidetti, R. and Jonsson, K.I. 2002. Long-term anhydrobiotic survival in semi-terrestrial micrometazoans. J Zool, London, 257: 181–187.

Guidetti, R., Boschini, D., Rebecchi, L. and Bertolani, R. 2006. Encystment processes and the "Matrioshka-like stage" in a moss-dwelling and in a limnic species of eutardigrades (Tardigrada). Hydrobiologia, 558: 9–21.

Guidi, L., Eitel, M., Cesarini, E. et al. 2011. Ultrastructural analyses support different morphological lineages in the Placozoa, Grell, 1971. J Morphol, 272: 371–378.

Gunathilaka, N., Niroshana, D., Amarasinghe, D. and Udayanga, L. 2018. Prevalence of gastrointestinal parasitic infections and assessment of deworming program among cattle and buffaloes in Gampha District, Sri Lanka. BioMed Res Int, Article Id: 3048373.

GUWS-Medical 2019. gumsmedical.inf/reproduction biology/rhombogen.hotmail.

Hafez, S.L. and Palanisamy, S. 2016. Nematodes associated with onion in Idaho and Eastern Oregon. University of Idaho. Bulletin 909: 1–8.

Hageman, G.S. 1983. A fine structural analysis of ovarian morphology, oogenesis, and ovulation in marine bryozoans *Membranipora serrilamella* (Cheilostomata, Anasca). PhD dissertation, University of Souther California.

Hagiwara, A. 1996. Appearance of floating resting eggs in the rotifers *Brachionus plicatilis* and *B. rotundiformis*. Bull Fac Fish, Nakasagi Univ, 77: 111–115.

Hajek, A.E. and Eilenberg, J. 2018. Biological control of invertebrate and vertebrate pests. In: *Natural Enemies: An Introduction to Biological Control*. (ed) Hejek, A.E., Cambridge University Press, pp 189–201.

Halbach, U. 1970. Die ursachen der temporal variation von *Brachionus calyciflorus* Pallas (Rotatoria). Oecologia, 4: 262–318.

Halbach, U. and Halbach-Keup, G. 1972. Einfluss von Aussenfaktosenauf den Fortplanzungsmodus heterogoner Rotatorien. Oecologia, 9: 203–214.

Halbach, U. 1973. Life table data and population dynamics of the rotifer *Brachionus calyciflorus* Pallus as influenced periodically oscillating temperatures. In: *Effects of Temperature on Ectothermic Organisms*. (ed) Wieser, W. Springer Verlag, Berlin, pp 217–228.

Hall, A. and Holland, C. 2000. Geographical variation in *Ascaris lumbricoides* fecundity and its implications for helminth control. Parasitol Today, 16: 540–544.

Haloti, S. 1993. Etude du parasite of du parasitisme: I'Orthonetide Intoshia liggniei endoparasite de I'Heteronemerte *Lineus ruber*. Ph.D. Thesis, Univ Reims, Champagne-Ardenne, p 133.

Hammond, P.M. 1992. Species Inventory. In: *Global Diversity, Status of the Earth's Living Resources*. (ed) Groombridge, B., Springer, Dordrecht, pp 17–39.

Hanelt, B. and Janovy, J. 1999. The life cycle of a horsehair worm, *Gordius robustus* (Nematomorpha: Gordioidea). J Parasitol, 85: 139–141.

Hanelt, B. 2009. An anomaly against a current paradigm–extremely low rates of individual fecundity variability of the gordian worm (Nematomorpha: Gordiida). Parasitology, 136: 211–218.

Hanelt, B., Bolek, M.G. and Schmidt-Rhaesa, A. 2012. Going solo: Discovery of the first parthenogenetic gordiid (Nematomorpha: Gordiida). PLoS ONE, 7: e34472.

Harmer, S.F. 1891a. On the British species of *Crisia*. J Cell Biol, 32: 127–182.

Harmer, S.F. 1891b. On the regeneration of lost parts in *Polyzoa*. Rep Brit Ass Adv Sci, 60: 862–863.

Harmer, S.F. 1913. The Polyzoa of waterworks. Proc Zool Soc Lond, 83: 426–457.

Hartfield, M. 2016. On the origin of asexual species by means of hybridization and drift. Mol Ecol, 25: 3264–3265.

Hartikinen, H. and Okamura, B. 2019. Ecology and evolution of malacosorean-bryozoan interactions. In: *Myxozoan Evolution, Ecology and Development*. (eds) Okamura B., Gruhl, A. and Bartholomew, J., Springer, Switzerland, pp 201–216.

Harzsch, S., Muller, C.H.G. and Perez, Y. 2015. Chaetognatha. In: *Evolutionary Developmental Biology of Invertebrates*. (ed) Wanninger, A., Springer Verlag, Wein. Vol 1, pp 215–240.

Hayward, P.J. and Ryland, J.S. 1973. Growth, reproduction and larval dispersal in *Alcyonidium hirsutum* (Fleming) and some other Bryozoa. Pub Staz Zool, Napoli, 39: 226–241.

Hedrick, L.R. 1935. The life history and morphology of *Spiroxys contortus* (Rudolphi); Nematoda: Spiruridae. Trans Am Microsc Soc, 54: 307–335.

Hejnol, A. 2015a. Acoelomorpha and Xenoturbellida. In: *Evolutionary Developmental Biology of Invertebrates*. (ed) Wanninger, A., Springer Verlag, Wein. Vol 1, pp 203–214.

Hejnol, A. 2015b. Gnathifera. In: *Evolutionary Developmental Biology of Invertebrates*. (ed) Wanninger, A., Springer Verlag, Wein. Vol 2. pp 1–12.

Hejnol, A. 2015c. Gastrotricha. In: *Evolutionary Developmental Biology of Invertebrates*. (ed) Wanninger, A., Springer Verlag, Wein. Vol 2, pp 13–20.

Hejnol, A. 2015d. Cycloneuralia. In: *Evolutionary Developmental Biology of Invertebrates: Ecdysozoa I: Non-Tetraconata*. (ed) Wanninger, A., Springer Verlag, Wein. Vol 3, pp 1–13.

Hema, P. and Khanna, A.S. 2018. Yield loss assessment of tomato through *Meloidogyne incognita* (kofoid and white) chitwood, in Himachal Pradesh, India. J Entomol Zool Stud, 6: 448–451.

Herbst, C. 1928. Untersuchungen zur Bestimmung des Geschlechts. I. Mitteilung: Ein neuer Weg zur Lösung des Geschlechtsbestimmungs-Problems bei *Bonellia viridis*. Sber Heidelb Akad Wiss, 19A: 1–19.

Hespeels, B., Knapen, M., Hanot-Mambres, D. et al. 2014. Gateway to genetic exchange? DNA double-stranded breaks in the bdelloid rotifer *Adineta vaga* submitted to desiccation. J Evol Biol, 27: 1334–1345.

Heupel, M.R. and Bennett, M.B. 1996. A myxosporean parasite (Myxosporea: Multivalvulidae) in the skeletal muscle of epaulette sharks, *Hemiscyllium ocellatum* (Bonnaterre), from the Great Barrier Reef. J Fish Dis, 19: 189–191.

Higgins, R.P. 1974. Kinorhyncha. In: *Reproduction of Marine Invertebrates*. (eds) Giese, A.C. and Pearse, J.S., Academic Press, New York, Vol 1, pp 507–518.

Higgins, R.P. 1986. A new species of *Echinoderes* (Kinorhyncha: Cyclorhagida) from a Coarse-sand California Beach. Trans Am Microsc Soc, 105: 266–273.

Higgins, R.P. and Storch, V. 1991. Evidence for direct development in *Meiopriapulus fijiensis* (Priapulida). Trans Am Microsc Soc, 110: 37–46.

Hino, A., Tanaka, T., Takaishi, M. et al. 2014. Karyotype and reproduction mode of the rodent parasite *Strongyloides venezuelensis*. Parasitology, 141: 1736–1745.

Hirayama, K., Takagi, K. and Kimura, H. 1979. Nutritional effect of eight species of marine phytoplankton on population growth of the rotifer, *Brachionus plicatilis*. Bull Jap Soc Sci Fish, 45: 11–16.

Hoeppli, R.J.C. 1926. Studies of free-living nematodes from the thermal waters of Yellowstone Park. Trans Am Microsc Soc, 45: 234–255.

Hogvall, M., Vellutini, B.C., Martin-Duran, J.M. et al. 2019. Embryonic expression of priapulid *Wnt* genes. Dev Genes Evol, 229: 125–135.

Holbein, J., Grundler, F.M.W. and Siddique, S. 2016. Plant basal resistance to nematodes: an update. J Exp Bot, DOI: 10.1093/jxb/erw005.

Holland, C. 2013. *Ascaris: The Neglected Parasites*. Newnes, p 460.

Hondt, J.L.D'. 2005. Les premiers bryozoologues et la connaissance des Bryozoaires de Rondelet a Linnaeus. Denisia, 16: 329–350.

Hope, W.D. 1974. Nematoda. In: *Reproduction of Marine Invertebrates: Acoelomate and Pseudocoelomate Metazoans*. (eds) Giese, A.C. and Pearse, J.S., Vol 1, Academic Press, New York, pp 391–469.

Horstadius, S. 1937. Experiments on determination in the early development of *Cerebratulus lacteus*. Biol Bull, 73: 317–342.

Hou, X., Wei, M., Li, Q. et al. 2019. Transcriptome analysis of larval segment formation and secondary loss in the echiuran worm *Urechis unicinctus*. Int J Mol Sci, 20: 1806.

Hughes, D.J. and Hughes, R.N. 1986. Life history variations in *Celleporella hyalina* (Bryozoa). Proc R Soc, 228B: 127–132.

Hughes, D.J. 1989. Variation in reproductive strategy among clones of the bryozoans *Celleporella hyaline*. Ecol Monogr, 59: 387–403.

Hughes, V., Benfey, T.J. and Martin-Robichaud, D.J. 2008. Effect of rearing temperature on sex ratio in juvenile Atlantic halibut, *Hippoglossus hippoglossus*. Env Biol Fish, 81: 415–419.

Hughes, R.N., Wright, P.J., Carvalho, G.R. and Hutchinson, W.F. 2009. Patterns of self compatibility, inbreeding depression, outcrossing, and sex allocation in a marine bryozoan suggest the predominating influence of sperm competition. Biol J Linn Soc, 98: 519–531.

Hugot, J.-P., Baujard, P. and Morand, S. 2001. Biodiversity in helminths and nematodes as a field of study: An overview. Nematology, 3: 1–10.

Humbert, J.-F. and Henry, C. 1989. Studies on the prevalence and the transmission of lung and stomach nematodes of the wild boar (*Sur scrofa*) in France. J Wildlife Dis, 25: 335–341.

Hummon, M.R. 1984. Reproduction and sexual development in a freshwater gastrotrich. 1. Oogenesis of parthenogenic eggs (Gastrotricha). Zoomorphology, 104: 33–41.

Hummon, M.R. 1986. Reproduction and sexual development in a freshwater gastrotrich. 4. Life history traits and the possibility of sexual reproduction. Trans Am Microsc Soc, 105: 97–109.

Hummon, W.D. 1974. Gastrotricha. In: *Reproduction of Marine Invertebrates*. (eds) Giese, A.C. and Pearse, J.S., Academic Press, New York, Vol 1, pp 485–506.

Hur, J.H., Doninck, K.V., Mandigo, M.L. and Meselson, M. 2009. Degenerate tetraploidy was established before bdelloid rotifer families diverged. Mol Biol Evol, 26: 375–383.

Huston, D.C., Cutmore, S.C. and Cribb, T.H. 2016. The life-cycle of *Gorgocephalus yaagi* Bray & Cribb, 2005 (Digenea: Gorgocephalidae) with a review of the first intermediate hosts for the superfamily Lepocreadioidea Odhner, 1905. Syst Parasitol, 93: 653–665.

Hutzell, P.A. and Krusberg, L.R. 1990. Temperature and the life cycle of *Heterodera zeae*. J Nematol, 22: 414–417.

Hyman, L.H. 1951a. *The Invertebrates: Platyhelminthes and Rhynchocoela*. McGraw-Hill Book, New York, Vol 2, p 550.

Hyman, L.H. 1951b. *The Invertebrates: Acanthocephala, Aschelminthes, and Entoprocta*. McGraw-Hill Book, New York, Vol 3, p 572.

Hyman, L.H. 1959. *The Invertebrates: Smaller Coelomate Groups*. McGraw-Hill Book, New York, Vol 5, p 783.

Ibrahim, M.M.I. 2008. Helminth infracommunities of the maculated toad *Amietophrynus regularis* (Anura: Bufonidae) from Ismailia, Egypt. Dis Aquat Org, 82: 19–26.

Inoue, I. 1962. Studies on the life history of *Chordodes japonensis*, a species of Gordiacea III. The modes of infection. Anno Zool Japan, 35: 12–19.

Iseto, T., Yokuta, Y. and Hirose, E. 2007. Seasonal change of species composition, abundance, and reproduction of solitary entoprocts in Okinawa Island, the Ryukyu Archipelago, Japan. Mar Biol, 151: 2099–2107.

Iseto, T. 2017. Review of the studies of Japanese entoprocts (Entoprocta). In: *Species Diversity of Animals in Japan. Diversity and Commonality in Animals*. (eds) Motokawa, M. and Kajihara, H., Springer, Tokyo, pp 445–467.

Iwata, F. 1958. On the development of the nemertean *Micrura akkeshiensis*. Embryologia, 4: 103–131.

Iwata, F. 1960. Studies on the comparative embryology of nemerteans with special reference to their interrelationships. Publ Akkeshi Mar Biol Stat, 10: 1–51.

Jagersten, G. 1964. On the morphology and reproduction of entoproct larvae. Zool Bidr Upsala, 36: 295–304.

Jakob, W., Sagasser, S., Dellaporta, S. et al. 2004. The *Trox-2Hox/Para-Hox* gene of *Trichoplax* (Placozoa) marks an epithelial boundary. Dev Genes Evol, 214: 170–175.

James, M.A., Ansell, A.d., Collins, M.J. et al. 1992. Biology of living brachiopods. Adv Mar Biol, 28: 175–387.

Jankowski, T., Collins, A.G. and Campbell, R. 2008. Global diversity of inland water cnidarians. Hydrobiologia, 595: 35–40.

Janssen, R., Wennberg, S.A. and Budd, G.E. 2009. The hatching larva of the priapulid worm *Halicryptus spinulosus*. Front Zool, 6: 8. DOI: 10.1186/1742-9994-6-8.

Janssen, T., Karssen, G., Topalovic, O. et al. 2017. Integrative taxonomy of root-knot nematodes reveals multiple independent origins of mitotic parthenogenesis. PLoS ONE, https://doi.org/10.1371/journal.pone.0172190.

Jarne, P. and Auld, J.R. 2006. Animals mix it up too: the distribution of self-fertilization among hermaphroditic animals. Evolution, 60: 1816–1824.

Jenner, R.A. 2014. Towards a phylogeny of the Metazoa: evaluating alternative phylogenetic positions of Platyhelminthes, Nemertea, and Gnathostomulida, with a critical reappraisal of cladistic characters. Contribution to Zoology, 73: 3–163.

Jennings, J.B. 1977. Patterns of nutritional physiology in free-living and symbiotic Turbellaria and their implications for the evolution of ectoparasitism in the phylum Platyhelminthes. Acta Zool Fennica, 154: 63–79.

Jensen, K. and Bullard, S.A. 2010. Characterization of diversity of tetraphyllidean and rhinebothriidean cestode larval types, with comments on host associations and life-cycles. Int J Parasitol, 40: 889–910.

Johnson, A.A. 1955. Life history studies on *Hydromermis contorta* (Kohn), a nematode parasite of *Chironomus plumosus* (L.) Ph.D. Thesis, University of Illinois, Urbana, USA.

Johnson, K.B. 2001. Sipuncula: The Peanut Worms. In: *An Identification Guide to the Larval Marine Invertebrates of the Pacific Northwest*. (ed) Shanks, A.C., Oregon State University Press, p 314.

Joilet, L. 1877. Contributions a l'histoire naturelle des bryozoaires des cotes de France. Arch Zool Exp Gen, 6: 193–304.

Jones, P.A. and Gilbert, J.J. 1977. Polymorphism and polyploidy in the rotifer *Asplanchna sieboldi*. Relative marker DNA contents in tissues of saccate and companulate females. J Exp Zool, 201: 163–168.

Jones, S.R.M., Prosperi-Porta, G., Dawe, S.C. and Barnes, D.P. 2003. Distribution, prevalence and severity of *Parvicapsula minibicornis* infections among anadromous salmonids in the Fraser River, British Columbia, Canada. Dis Aquat Org, 54: 49–54.

Jonsson, K.I. and Bertolani, R. 2001. Facts and fiction about long-term survival in tardigrades. J Zool, 255: 121–124.

Jonsson, K.I. and Rebecchi, L. 2002. Experimentally induced anhydrobiosis in the tardigrade *Richtersius coronifer*: Phenotypic factors affecting survival. J Exp Zool, 293: 578–584.

Jung, K.J., Woss, E.R., Chae, H.S. and Seo, J.E. 2017. Gymnolaemate bryozoans in fresh and brackish water of South Korea: occurrence, taxonomical remarks and zoogeographical implications. Anim Syst Evol Divers, 33: 37–44.

Kallert, D.M. 2006. Life cycle studies and transmission mechanisms of myxozoan parasites. Ph.D. Thesis, Friedrich-Alexander University, Erlangen-Nurnberg, Germany.

Kar, A., Mondal, S. and Chakraborty, G. 2018. Management of root knot nematode (*Meloidogyne incognita*) race 2 in cowpea through bio-agents. J Crop Weed, 14: 220–223.

Karlson, R.H. 1991. Recruitment and local persistence of a freshwater bryozoans in stream riffles. Hydrobiologia, 226: 119–128.

Kaulfuss, A., Seidel, R. and Luter, C. 2013. Linking micromorphism, brooding, and hermaphroditism in brachiopods: insight from Caribbean *Argyrotheca* (Brachiopoda). J Morphol, 274: 361–376.

Kaur, H. 2014. Myxozoan infestation in freshwater fishes in wetlands and aquaculture in Punjab (India). Adv Anim Vet Sci, 2: 488–502.

Kelehear, C., Spratt, D.M., O'Meally, D. and Shine, R. 2014. Pentastomids of wild snakes in the Australian tropics. Int J Parasitol: Parasite & Wildlife, 3: 20–31.

Kennedy, C.R. 1993. Acanthocephala. In: *Asexual Reproduction and Reproductive Biology of Invertebrates*. (eds) Adiyodi, A.D. and Adiyodi, K.G., Vol 6A, Oxford and IBH Publishers, New Delhi, pp 279–295.

Keymer, A., Crompton, D.W.T. and Walters, D.E. 1983. Parasitic population ecology and host nutrition. Dietary fructose and *Moniliformis* (Acanthocephala). Parasitology, 87: 265–278.

Kimble, J. and Ward, S. 1988. Germ-line development and fertilization. In: *The Nematode Caenorhabditis elegans.* (ed) Wood, W.B., Cold Spring Harbor Laboratory Press, pp 191–213.

King, C.E. 1970. Comparative survivorship and fecundity of mictic and amictic female rotifers. Physiol Zool, 43: 206–212.

King, C.E. and Miracle, M.R. 1980. A perspective on aging in rotifers. Hydrobiologia, 73: 13–19.

King, C.E. 1983. A re-examination of Lausing Effect. Hydrobiologia, 104: 135–139.

Kirankumar, S. and Pandian, T.J. 2003. Production of androgenetic tiger barb, *Puntius tetrazona.* Aquaculture, 228: 37–51.

Kirk, K.L. 1997. Life-history responses to variable environments: Starvation and reproduction in planktonic rotifers. Ecology, 78: 434–441.

Klass, M.R. 1977. Aging in the nematode *Caenorhabditis elegans*: major biological and environmental factors influencing life span. Mech Age Dev, 6: 413–429.

Kobylinski, K.C., Sylla, M., Black IV, W. and Foy, B.D. 2012. Mermithid nematodes found in adult *Anopheles* from southeastern Senegal. Parasites & Vectors, 5: 131.

Kohler, S.L. 2008. The ecology of host-parasite interactions in aquatic insects. In: *Aquatic Insects.* (eds) Lancaster, J. and Briers, R.A., CABI International, London, pp 55–80.

Kolasa, J. and Tyler, S. 2010. Flatworms: Turbelarians and Nemertea. In: *Ecology and Classification of North American Freshwater Invertebrates.* (eds) Thorp, J.H. and Covich, A.P., Elsevier, Amsterdam, pp 143–161.

Konno, K. 1971. On some entoprocts found at Fukaura, Aomori Prefecture. Rep Fukaura Mar Biol, 3: 2–9.

Koprivnikar, J. and Randhawa, H.S. 2013. Benefits of fidelity: does host specialization impact nematode parasite life history and fecundity? Parasitology, 140: 587–597.

Kozloff, E.N. 1965. *Ciliocincta sabellariae* gen and sp, n., an orthonectid mesozoan from the polychaete *Sabellaria cementarium*, Moore. J Parasitol, 51: 37–44.

Kozloff, E.N. 1971. Morphology of the orthonectid *Ciliocincta sabellariae.* J Parasitol, 57: 585–597.

Kozloff, E.N. 1972. Some aspects of development in *Echinoderes* (Kinorhyncha). Trans Am Microsc Soc, 91: 119–130.

Kozloff, E.N. 1997. Studies on the so-called plasmodium of *Ciliocinta sabellariae* (Phylum Orthonectida), with notes on an associated microsporan parasite. Cah Biol Mar, 38: 151–159.

Kozloff, E.N. 2007. Stages of development from first cleavage to hatching, of an *Echinoderes* (Phylum Kinorhyncha: Class Cyclorhagida). Cah Biol Mar, 48: 199–206.

Kristensen, R.M. 1983. Loricifera, a new phylum with Aschelminthes characters from the meiobenthos. Z Zool Syst Evol Forsch, 21: 163–180.

Kristensen, R.M. 1995. Are aschelminthes pseudocoelomate or acoelomate? In: *Body Cavities Function and Phylogeny.* (eds) Lacivelli, G., Valvassori, R. and Carnevali, M.D.C., Symp Monogr, pp 41–43.

Kristensen, R.M. 2002. An introduction to Loricifera, Cycliophora, and Micrognathozoa. Integ Comp Biol, 42: 641–651.

Kubecka, B., Bruno, A. and Rollins, D. 2017. Geographic survey of *Oxyspirura petrowi* among wild Northern Bobwhites in the United States. Natl Quail Symp Proc, 8: 311–315.

Kuris, A.M. 1993. Life cycles of nemerteans that are symbiotic egg predators of decapods Crustacea: adaptations to host life histories. Hydrobiologia, 266: 1–14.

Lancaster, J. and Bovil, W.D. 2017. Species-specific prevalence of mermithid parasites in populations of six congeneric host caddisflies of *Ecnomus* McLachlan, 1864 (Trichoptera: Ecnomidae). Aqua Insects, https://dx.doi.org/10.1080/01650421.2017.1299866.

Land, J.v.d. 1970. Systematics, zoogeography, and ecology of the Priapulida. Zool Verh, 112: 1–118.

Land, J.v.d. 1975. Priapulida. In: *Reproduction of Marine Invertebrates.* (eds) Giese, A.C. and Pearse, J.S., Academic Press, New York, Vol 2, pp 55–66.

Land, van der J. 1968. A new aschelminth, probably related to the Priapulida. Zool Meded, Leiden, 42: 237–250.

Lapan, E.A. and Morowitz, H.J. 1972. The Mesozoa. Sci Amer, 222: 94–101.

Laws, A.N. 2009. Density-dependent reductions in grasshopper fecundity in response to nematode parasitism. Can Entomol, 141: 415–421.

Lefevre, G. 1907. Artificial parthenogenesis in *Thalassema mellita*. J Exp Zool, 4: 91–149.

Lemaire, 2011. Evolutionary crossroads in developmental biology: the tunicates. Development, 138: 2143–2152.

Leutert, R. 1974. Zur Geschelechtsbestimmung und Gametogenese von *Benellia viridis* Rolando. J Embryol Exp Morph, 32: 169–193.

Li, J., Todd, T.C., Oakley, T.R. et al. 2010. Host-derived suppression of nematode reproductive and fitness genes decreases fecundity of *Heterodera glycines* Ichinohe. Planta, 232: 775–785.

Littlewood, D.T.J., Bray, R.A. and Waeschenbach, A. 2015. Phylogenetic patterns of diversity in cestodes and trematodes. In: *Parasite Diversity and Diversification: Evolutionary Ecology Meets Phylogenetics*. (eds) Morand, S., Krasnov, B.R. and Littlewood, D.T.J., Cambridge University Press, pp 304–319.

Liu, Q.L., Thomas, V.P. and Williamson, V.M. 2007. Meiotic parthenogenesis in a root-knot nematode results in rapid genomic homozygosity. Genetics, 176: 1483–1490.

Lom, J. and Dykova, I. 2006. Myxozoan genera: definition and notes on taxonomy, life-cycle terminology and pathogenic species. Folia Parasitol, 53: 1–36.

Lopez-Legentil, S., Erwin, P.M., Velasco, M. and Turon, X. 2013. Growing or reproducing in a temperate sea: optimization of resource allocation in a colonial ascidian. Invert Biol, 132: 69–80.

Lovell, J.T., Aliyu, O.M., Mau, M. et al. 2013. On the origin and evolution of apomixes in *Boechera*. Plant Reprod, 26: 309–315.

Lubzens, E., Tandler, A. and Minkoff, G. 1989. Rotifer as food in aquaculture. Hydrobiologia, 186-187: 387–400.

MacGinitie, G.E. 1935. Normal functioning and experimental behaviour of the egg and sperm collectors of the echiuroid, *Urechis caupo*. J Exp Zool, 70: 341–355.

Maiorova, A.S. and Adrianov, A.V. 2017. Deep-sea sipunculans (Sipuncula) of the northwestern Pacific. Rus J Mar Biol, 43: 181–189.

Mano, R. 1960. On the metamorphosis of a brachiopod, *Frenulina sanguinolenta* (Gmelin). Bull Mar Biol Stat Asamushi, 10: 171–175.

Manriquez, P.H. and Cancino, J.M. 1996. Bryozoan-macroalgal interactions: do epibionts benefit? Mar Evol Prog Ser, 138: 189–197.

Manriquez, P.H., Hughes, R.N. and Bishop, J.D.D. 2001. Age-dependent loss of fertility in water-borne sperm of the bryozoans *Celleporella hyalina*. Mar Ecol Prog Ser, 224: 87–92.

Marcus, E. 1934. Uber *Lophopus crystallinus*. Zool Jahrb Abt Ant, 58: 501–606.

Marcus, E. 1939. Bryozoarios marinhos brasileiros III. Bolm Fac filos Cienc Univ S Paulo Zool, 3: 111–354.

Mariscal, R.N. 1975. Entoprocta. In: *Reproduction of Marine Invertebrates*. (eds) Giese, A.C. and Pearse, J.S., Academic Press, New York, Vol 2, pp 1–42.

Marsden, J.C.R. 1957. Regeneration in *Phoronis vancouverensis*. J Morphol, 101: 307–323.

Martindale, M.Q. and Henry, J.Q. 1995. Modifications of cell fate specification in equal-cleaving nemertean embryos: alternate patterns of spiralian development. Development, 121: 3175–3185.

Martin-Duran, J.M., Janssen, R., Wennberg, S. et al. 2012. Deuterostomic development in the protostome *Priapulus caudatus*. Curr Biol, 22: 2161–2166.

Martin-Duran, J.M., Vellutini, B.C. and Hejnol, A. 2015. Evolution and development of the adelphophagic, intracapsular Schmidt's larva of the nemertean *Lineus ruber*. Evol Dev, 6: 28, DOI: 10.1186/s13227-015-0023-5.

Martins, M.L., Ghiraldelli, L., Garcia, F. et al. 2007a. Experimental infection in *Notodiaptomus* sp (Crustacea: Calanoida) with larvae of *Cammalanus* sp (Nematoda: Cammalanidae). Arq Bras Med Vet Zootec, 59: 382–386.

Martins, M.L., Garcia, F., Piazza, R.S. and Ghiradelli, L. 2007b. *Camallanus maculatus* n sp (Nematoda: Cammalanidae) in an ornamental fish *Xiphophorus maculatus* (Osteichthyes: Poeciliidae) cultivated in Sao Paulo State, Brazil. Arq Bras Med Vet Zootec, 59: 1224–1230.

Maslakova, S. and von Dohren, J. 2009. Larval development with transitory epidermis inn *Paranemertes peregrina* and other loplonemerteans. Biol Bull, 216: 273–292.

Maslakova, S.A. and Hiebert, T.C. 2014. From trochophore to pilidium and back again–a larva's journey. Int J Dev Biol, 58: 585–591.

Massard, J.A. and Geimer, G. 2008. Global diversity of bryozoans (Bryozoa or Ectoprocta) in freshwater: an update. Bull Soc Nat Luxemb, 109: 139–148.

Mavrot, F., Hertzberg, H. and Torgerson, P. 2015. Effects of gastro-intestinal nematode infection on sheep performance: a systematic review and meta-analysis. Parasites & Vector, 8: 557, DOI: 10.1186/s13071-015-1164-z.

Mayer, G., Franke, F.A., Treffkorn, S. et al. 2015. Onychophora. In: *Evolutionary Developmental of Invertebrates: Ecdysozoa I: Non-Tetraconata.* (ed) Wanninger, A., Springer Verlag, Wein. pp 53–98.

McConnaughey, B.H. 1989. Mesozoa. In: *Reproductive Biology of Invertebrates.* (eds) Adiyodi, K.G. and Adiyodi, R.G., Oxford and IBH Publishers, New Delhi, Vol 4A: pp 135–146.

McConnaughey, B.H. 1993. Mesozoa. In: *Reproductive Biology of Invertebrates.* (eds) Adiyodi, K.G. and Adiyodi, R.G., Oxford and IBH Publishers, New Delhi, Vol 6A: pp 219–230.

McGruk, C., Morris, D.J., Bron, J.E. and Adams, A. 2005. The morphology of *Tetracapsuloides bryosalmonae* (Myxozoa: Myxosporea) spores released from *Fredericella sultana* (Bryozoa: Phylactolaemata). J Fish Dis, 28: 1–6.

Meadow, P.S., Reichelt, A.C., Meadow, A. and Waterworth, J.S. 1994. Microbial and meiofaunal abundance, redox potential, pH and shear strength profiles in deep sea Pacific sediments. J Geol Soc, 151: 377–390.

Medeiros, J.F., Py-Daniel, V., Barbosa, U.C. and Izzo, T.J. 2009. *Mansonella ozzardi* in Brazil: prevalence of infection in riverine communities in the Purus regions, in the state of Amazonas. Mem Inst Oswaldo Cruz, 104: 74–80.

Mehlhorn, H. 2016. *Encyclopedia of Parasitology.* Springer Verlag, New York, p 1573.

Meidlinger, K., Tyler, P.A. and Peck, L.S. 1998. Reproductive patterns in the Antarctic brachiopod *Liothyrella uva.* Mar Biol, 132: 153–162.

Mendes, F., De Figueiredo, G.M. and Valentin, J.L. 2012. Reproduction and structure of the population of the Chaetognath *Parasagitta friderici* in Guanabara Bay (Brazil) based on short term sampling. Ann Brazil Acad Sci, 84: 103–111.

Menon, N.R. 1972. Heat tolerance, growth and regeneration in three North Sea bryozoans exposed to different constant temperatures. Mar Biol, 15: 1–11.

Meyl, A.H. 1954. Beitrage zur Kenntnis der Nematodenfauna vulkanisch erhitzter Biotope. Z Morphol Okol Tiere, 42: 421–448.

Milanin, T., Eiras, J.C., Arana, S. et al. 2010. Phylogeny, ultrastructure, histopathology and prevalence of *Myxobolus oliveirai* sp nov, a parasite of *Brycon hilarii* (Characidae) in the Pantanal wetland, Brazil. Mem Inst Oswaldo Cruz, 105: 762–769.

Miller, H.M. 1931. Alternation of generations in the rotifer *Lecane inermis* Bryce. Biol Bull, 60: 345–381.

Miller, W.R. 1997. Tardigrades: Bears of the moss. Kans School Nat, 43: 3–15.

Monge-Najera, J. 1994. Reproductive trends, habitat type and body characteristics in velvet worms (Onychophora). Rev Biol Trop, 42L3: 611–622.

Monteiro, C.M., Amato, J.F.R. and Amato, S.B. 2006. A new species of *Andracantha* Schmidt (Acanthocephala, Polymorphidae) parasite of Neotropical cormorants, *Phalacrocorax brasilianus* (Gmelin) (Aves, Phalacrocoracidae) from Southern Brazil. Revis Brasil Zool, 23: 807–812.

Moorthy, V.N. 1938. Freshwater nematodes from the intestine of fish. Proc Helm Soc Wash, 5: 24–28.

Morand, S. 1996. Life-history traits in parasitic nematodes: a comparative approach for the search of invariants. Funct Ecol, 10: 210–218.

Morand, S., Bouamer, S. and Hugot, J-P. 2005. Nematodes. In: *Micromammals and Macroparasites: From Evolutionary Ecology to Management.* (eds) Morand, S., Krasinov, B. and Poulin, R., Springer-Verlag, Berlin, pp 63–80.

Moreira, R.A., Mansano, A.S. and Rocha, O. 2016. Life cycle traits of *Philodina roseola* Ehrenberg, 1830 (Rotifera, Bdelloidea), a model organism for bioassays. An Acad Bras Cienc, 88: 579–588.

Morris, D.J. and Adams, A. 2008. Sporogony of *Tetracapsuloides bryosalmonae* in the brown trout *Salmo trutta* and the role of the tertiary cell during the vertebrate phase of myxozoan life cycles. Parasitology, 135: 1075–1092.

Muazu, L., Abdullahi, Y. and Umar, Z. 2017. Prevalence of human intestinal parasitic nematode among outpatients attending Wudil General Hospital, Kano State, Nigeria. Int J Pharmaceutical Clinic Res, 9: 247–251.

Mukai, H. and Makioka, T. 1980. Some observations on the sex differentiation of an entoproct, *Barentsia discreta* (Busk). J Exp Zool, 213: 45–59.

Muller, M., Mental, M., van Hellemond, J.J. et al. 2012. Biochemistry and evolution of anaerobic energy metabolism in eukaryotes. Microbiol Mol Biol, 76: 444–495.

Murugesan, P., Balasubramanian, T. and Pandian, T.J. 2010. Does haemocoelom exclude embryonic stem cells and asexual reproduction in invertebrates? Curr Sci, 98: 768–770.

Nasanov, N.V. 1926. *Arthropodaria kovalevskii* n sp (Entoprocta) und die Regeneration ihrer Organe. Wiss Arbeite der zoologischen Laboratoriums der biologischen Station Akad Wiss SSSR, Sebastopol (Ser. II), 5: 1–38.

Near, T.J. 2002. Acanthocephalan phylogeny and the evolution of parasitism. Integ Comp Biol, 42: 668–677.

Nekliudova, U.A., Shunkina, K.V., Grishankov, A.V. et al. 2019. Colonies are dynamic systems: reconstructing the life history of *Cribrilina annulata* (Bryozoa) on two algal substrates. J Mar Biol Ass UK, https://doi.org/10.1017/S0085315419000286.

Nelson, D.R., Guidetti, R. and Rebecchi, L. 2010. Tardigrada. In: *Ecology and Classification of North American Freshwater Invertebrates*. (eds) Thorp, J.H. and Covich, A.P., Elsevier, Amsterdam, pp 455–484.

Nesteruk, T. 1991. Vertical distribution of Gastrotricha in organic bottom sediment of inland water bodies. Acta Hydrobiol, 33: 253–264.

Neuhaus, B. and Higgins, R.P. 2002. Ultrastructure, biology, and phylogenetic relationships of Kinorhyncha. Integ Comp Biol, 42: 619–632.

Neves, R.C., Gambi, C., Danovaro, R. and Kristensen, R.M. 2014. *Spinoloricus cinziae* (Phylum Loricifera), a new species from a hypersaline anoxic deep basin in the Mediterranean Sea. System Biodiv, DOI: 10.1080/14772000.2014.943820.

Neves, R.C. and Reichert, H. 2015. Microanatomy and development of the dwarf male of *Symbion pandora* (Phylum Cycliophora): New insights from ultrastructural investigation based on serial section electron microscopy. PLoS ONE, 10: e0122364.

Newmann, S., Reuner, A., Brummer, F. and Schill, R.O. 2009. DNA damage in storage cells of anhydrobiotic tardigrades. Comp Biochem Physiol A, 153: 425–429.

Nicholas, W.L. and Hynes, H.B. 1958. Studies on *Polymorphus minutus* (Goeze, 1782) (Acanthocephala) as a parasite of the domestic duck. Ann Trop Med Parasitol, 52: 36–47.

Nicholas, W.L. and Hynes, H.B.N. 1963. The embryology of *Polymorphus minutus* (Acanthocephala). Proc Zool Soc London, 141: 791–801.

Nickol, B.B. 2006. Phylum Acanthocephala. In: *Fish Diseases and Disorders*. (ed) Woo, P.T.K., CABI International, UK. Vol 1: 444–452.

Nielsen, C. 1964. Studies on Danish Entoprocta. Ophelia, 1: 1–76.

Nielsen, C. 1966. Some Loxosomatidae (Entoprocta) from the Atlantic coast of the United States. Ophelia, 3: 249–275.

Nielsen, C. 1990. Bryozoa Entoprocta. In: *Reproductive Biology of Invertebrates*: *Fertilization, Development, and Parental Care*. (eds) Adiyodi, K.G and Adiyodi, R.G., Vol 4, Part B. Oxford and IBH Publishers, New Delhi, pp 201–210.

Nielsen, C. 1991. The development of the brachiopod *Crania* (*Neocrania*) *anomala* (O.F. Muller) and its phylogenetic significance. Acta Zool, 72: 7–28.

Nielsen, C. 2013. Entoprocta. In: *Introduction to the Study of Meiofauna*. (eds) Higgins, R.P. and Thiel, H., Smithsonian Institution Press, pp 444.

Nigon, V. 1965. Developpement et reproduction des Nematodes. In: *Traite de Zoologie*. (ed) Grasse, P.P., Masson, Paris, 4: 218–386.

Nusbaum, J. and Oxner, M. 1913. Die Embryonalentwicklung des *Lineus ruber*. Zeitsch Wissen Zool, 107: 78–191.

O'Connor, R.J., Seed, R. and Boaden, P.J.S. 1980. Resource space partitioning by the Bryozoa of the *Fucus serratus* L. Comm. J Exp Mar Biol Ecol, 45: 117–137.

O'Dea, A. 2006. Asexual propagation in the marine bryozoans *Cupuladria exfragminis*. J Exp Mar Biol Ecol, 335: 312–322.

Oka, H. and Watanabe, H. 1957. Colony-specificity in compound ascidians. Bull Mar Biol Stn Asamushi, 10: 153–155.

Okamura, B. 1996. Occurrence, prevalence, and effects of the myxozoan *Tetracapsula bryozoides* parasitic in the freshwater bryozoans *Cristatella mucedo* (Bryozoa: Phylactolaemata). Folia Parasitol, 43: 262–266.

Okamura, B., Gruhl, A. and Bartholomew, J.L. 2015. *Myxozoan Evolution, Ecology and Development*. Springer International, Switzerland, p 441.

Okamura, B. and Gruhl, A. 2016. Myxozoa + Polypodium: A common route to endoparasitism. Trends Parasitol, 32: 268–271.

Okamura, B., Hartikainen, H. and Trew, J. 2019. Waterbird-mediated dispersal and freshwater biodiversity: general insights from bryozoans. Front Ecol Evol, http://doi.org/10.3389/fevo.2019.00029.

Olivier, J. 1966. Cytochimie de l'Ovocyte au Cours de la vitellogenese chez *Lineus ruber* (Nemerte). Ann Univ Reims L'Arers, 4: 158–165.

Ostrovsky, A.N. 2013. *Evolution of Sexual Reproduction in Marine Invertebrates*. Springer-Verlag, Vienna, p 356.

Otto, F. 1921. Regulationsvermogen einiger Susswasserbryozoen. Arch Entw Organ, 47: 205–208.

Pai, S. 1927. Lebenzyklus der *Anguillula aceti*. Ehrbg Zool Anz, 74: 257–270.

Pandian, T.J. 1975. Mechanism of heterotrophy. In: *Marine Ecology*. (ed) Kinne, O., John Wiley, London, 3 Part 1: 61–249.

Pandian, T.J. 2010. *Sexuality in Fishes*. Science Publishers/CRC Press, USA, p 208.

Pandian, T.J. 2012. *Sex Determination in Fish*. Science Publishers/CRC Press, USA, p 270.

Pandian, T.J. 2014. *Endocrine Sex Differentiation in Fish*. CRC Press, USA, p 303.

Pandian, T.J. 2015. *Environmental Sex Determination in Fish*. CRC Press, USA, p 299.

Pandian, T.J. 2016. *Reproduction and Development in Crustacea*. CRC Press, USA, p 301.

Pandian, T.J. 2017. *Reproduction and Development in Mollusca*. CRC Press, USA, p 299.

Pandian, T.J. 2018. *Reproduction and Development in Echinodermata and Prochordata*. CRC Press, USA, p 270.

Pandian, T.J. 2019. *Reproduction and Development in Annelida*. CRC Press, USA, p 276.

Pandian, T.J. 2020. *Reproduction and Development in Platyhelminthes*. CRC Press, USA, p 303.

Pandian, T.J. 2021. *Evolution and Speciation in Animals (in preparation)*.

Parenti, U., Antoniotti, M.L. and Beccio, C. 1965. Sex ratio and sex digamety in *Echinorhynchus truttae*. Experoemtoa. 21: 657–658.

Parshad, V.R. and Guraya, S.S. 1978. Morphological and histochemical observations on oocyte atresia in *Centrorhynchus corvi* (Acanthocephala). Parasitology, 77: 133–138.

Parshad, V.R. and Crompton, D.W. 1981. Aspects of acanthocephalan reproduction. Adv Parasitol, 19: 73–138.

Paul, M. 1970. Fertilization-associated changes in the eggs of *Strongylocentrotus purpuratus* and *Urechis caupo*. Doctoral dissertation, Standford University, Standford, California.

Pawlowski, J., Montoya-Burgos, J.I., Fahrni, J.F. et al. 1996. Origin of the Mesozoa inferred from 18S rRNA gene sequences. Mol Biol Evol, 13: 1128–1132.

Peebles, F. and Fox, D. 1933. The structure, functions and general reactions of *Dendrostoma zostericola*. Bull Scripps Inst Oceanogr, Univ California Tech Ser, 3: 201–224.

Pemberton, A.J., Hughes, R.N., Manriques, P.H. and Bishop, J.D.D. 2003. Efficient utilization of very dilute aquatic sperm: sperm competition may be more likely than sperm limitation when eggs are retained. Proc R Soc B, 270S: 223–226.

Pemberton, A.J., Hansson, L.J., Craig, S.F. et al. 2007. Microscale genetic differentiation in a sessile invertebrate with cloned larvae: investigating the role of polyembryony. Mar Biol, 153: 71–82.

Perez, Y., Muller, C.H.G. and Harzsch, S. 2014. The Chaetognatha: an anarchistic taxon between Protostomia and Deuterostomia. In: *Deep Metazoan Phylogeny: The Backbone of the Tree of Life*. (eds) Wagele, J.W. and Bartolomaeus, T., Walter De Gruyter, Berlin, pp 49–74.

Permin, A., Bisgaard, M., Frandsen, F. et al. 1999. Prevalence of gastrointestinal helminthes in different poultry productions systems. British Poultry Sci, 40: 439–443.

Petersen, J.J. 1975. Development and fecundity of *Reesimermis nielseni*, a nematode parasite of mosquitos. J Nematol, 7: 211–214.

Pilarska, J. 1972. The dynamics of growth of experimental populations of the rotifer *Brachionus rubens* Ehrbg. Pol Arch Hydrobiol, 19: 265–277.

Pilger, J. 1978. Settlement and metamorphosis in the Echiura: A review. In: *Settlement and Metamorphosis of Marine Invertebrate Larvae.* (eds) Chia, F.-S. and Rice, M.E., Elsevier, North-Holland Biomedical Press, pp 103–112.

Pilger, J.F. 1978. Reproduction of a parthenogenetic sipunculan. Amer Zool, 18: 663.

Pilger, J.F. 1987. Reproductive biology and development of *Themiste lageniformis*. A parthenogenic sipunculan. Bull Mar Sci, 41: 59–67.

Poinar, G. Jr. 2008. Global diversity of hairworms (Nematomorpha: Gordiaceae) in freshwater. Hydrobiologia, 595: 79–83.

Poinar, G. Jr. 2010 Nematoda and Nematomorpha: In: *Ecology and Classification of North American Freshwater Invertebrates.* (eds) Thorp, J.H. and Covich, A.P., Elsevier, Amsterdam, pp 237–280.

Poinar, G.O. Jr. and Gyrisco, G.G. 1962. A new mermithid parasite of the alfalfa weevil, *Hypera postica* (Gyllenhal). J Insect Pathol, 4: 201–206.

Poinar, G.O. Jr., Lane, R.S. and Thomas, G.M. 1976. Biology and redescription of *Pheromermis pachysoma* (V. Linstow) N. Gen., N. Comb. (Nematoda: Mermithidae), a parasite of yellowjackets (Hymenoptera: Vespidae). Nematologica, 22: 360–370.

Pollock, L.W. 1975. Tardigrada. In: *Reproduction of Marine Invertebrates.* (eds) Giese, A.C. and Perase, J.S., Academic Press, New York, Vol 2, pp 43–54.

Potts, F.A. 1910. Notes on the free-living nematodes. J Cell Sci, 55: 433–484.

Poulin, R. and Morand, S. 2000. The diversity of parasites. Q Rev Biol, 75: 277–293.

Poulin, R. and Latham, A.D.M. 2002. Inequalities in size and intensity-dependent growth in a mermithid nematode parasitic in beach hoppers. J Helminthol, 76: 65–70.

Pourriot, R. and Rieunier, M. 1973. Recherches sur la biologie des rotifers. III. Fecondite et duree de vie compare chez les femelles amictiques et mictiques de quelque especes. Annls Limnol, 9: 241–258.

Prots, P., Wanninger, A. and Schwaha, T. 2019. Life in a tube: morphology of the ctenostome bryozoans *Hypophorella expansa*. Zool Lett, 5: 28, https://doi.org/10.1186/240851-019-0142-2.

Rahm, G. 1937. Grenzen des Lebens? Studien in heissen Quellen. Forsch Fortsch, 13: 381–387.

Rahm, P.G. 1923. Biologische und physiologische Beitrage zur Kenntnis der Moosfauna. Z Allg Physiol, 20: 1–34.

Ramudu, K.R. and Dash, G. 2016. Prevalence, morphology and scanning electron microscopy study of myxozoan parasites. J Parasit Dis, 40: 339–347.

Rao, D.G. and Satapathy, S. 1996. Demecology of Kinorhyncha of Chilka lagoon, Bay of Bengal. J Mar Biol Ass India, 38: 15–24.

Rao, T.R. and Sarma, S.S.S. 1985. Mictic and amictic modes of reproduction in the rotifer *Brachionus patulus* Mueller. Curr Sci, 54: 499–500.

Ravinet, N., Chartier, C., Bareille, N. et al. 2016. Unexpected decrease in milk production after Fenbendazole treatment of dairy cows during early grazing season. PLoS ONE, 11(1): e0147835, https://doi.org/10.1371/journal.pone.0147835.

Rebecchi, L., Boschini, D., Cesari, M. et al. 2009a. Stress response of a boreoalpine species of tardigrade, *Borealibius zetlandicus* (Eutardigrada, Hypsibiidae). J Limnol, 68: 64–70.

Rebecchi, L., Cesari, M. Altiero, T. et al. 2009b. Survival and DNA degradation in anhydrobiotic tardigrades. J Exp Biol, 212: 4033–4039.

Reeve, M.R. 1970. The biology of Chaetognatha. I. Quantitative aspects of growth and egg production in *Sagitta hispida*. In: *Marine Food Chains.* (ed) Steele, J.H., Oliver and Boyd, Edinbourgh, pp 168–189.

Reeve, M.R. and Cosper, T.C. 1975. Chaetognatha. In: *Reproduction of Marine Invertebrates.* (eds) Giese, A.C. and Pearse, J.S., Academic Press, New York. Vol 2, pp 157–184.

Reuner, A., Hengherr, S., Brummer, F. and Schill, R.O. 2010. Comparative studies on storage cells in tardigrades during starvation and anhydrobiosis. Curr Zool, 56: 259–263.

Ricci, C., Melone, G. and Sotgia, C. 1993. Old and new data on Seisonidea (Rotifera). Hydrobiologia, 255/256: 495–511.

Ricci, C. 1995. Growth pattern of four strains of a Bdelloid rotifer species: egg size and numbers. Hydrobiologia, 313-314: 157–163.

Ricci, C. and Fasico, U. 1995. Life-history consequences of resource allocation of two bdelloid rotifer species. Hydrobiologia, 299: 231–239.

Rice, M.E. 1970. Asexual reproduction in a sipunculan worm. Science, 167: 1618–1620.

Rice, M.E. 1975. Sipuncula. In: *Reproduction of Marine Invertebrates*. (eds) Giese, A.C. and Pearse, J.S., Academic Press, New York, Vol 2, pp 67–128.

Rice, M.E. 1981. Larvae adrift: patterns and problems in life histories of sipunculans. Am Zool, 29: 605–619.

Rice, M.E. 1989. Sipuncula. In: *Reproductive Biology of Invertebrates*. (eds) Giese, A.C. and Pearse, J.S., Oxford & IBH, New Delhi, Vol 4A, pp 263–280.

Rice, M.E., Reichardt, H.F., Piraino, J. and Young, C.M. 2012. Reproduction, development, growth, and the length of larval life of *Phascolosoma turnerae*, a wood-dwelling deep-sea sipunculan. Invert Biol, 10: 1–12.

Rico-Martinez, R. and Walsh, E.J. 2013. Sexual reproductive biology of a colonial rotifer *Sinantherina socialis* (Rotifera: Monogononta): Do mating strategies vary between colonial and solitary rotifer species? Mar Freshw Behav Physiol, 46: 419–430.

Riley, J. 1981. An experimental investigation of the development of *Porocephalus crotali* (Pentastomida: Porocephalida) in the western diamondback rattlesnake (*Crotalus artox*). Int J Parasitol, 11: 127–131.

Riser, N.W. 1974. Nemertinea. In: *Reproduction of Marine Invertebrates: Acoelomate and Pseudocoelomate Metazoans*. (eds) Giese, A.C. and Pearse, J.S., Academic Press, New York, 1: 359–390D.

Ritzmann, N.F., da Rocha, R.M. and Roper, J.J. 2009. Sexual and asexual reproduction in *Diademnum rodriguesi* (Ascidiacea, Didemnidae). Iheringia Ser Zool Porto Alegre, 99: 106–110.

Robinson, E.S. 1965. The chromosomes of *Moniliformis dubius* (Acanthocephala). J Parasitol, 51: 430–432.

Rodriguez-Vivas, R.I., Grisi, L., de Leon, A.A.P. and Villela, H. 2017. Potential economic impact assessment for cattle parasites in Mexico. Rev Mex Cien Pec, 8: 61–74.

Rougier, C., Pourriot, R. and Lam-Hoai, T. 2000. The genus *Synchaeta* (rotifers) in a northwestern Mediterranean coastal lagoon (Etang de Thau, France): taxonomical and ecological remarks. Hydrobiologia, 436: 105–117.

Ruhberg, H. 1990. Onychophora. In: *Reproductive Biology of Invertebrates*. (eds) Adiyodi, K.G. and Adiyodi, R.G., Oxfored & IBH Publishers, New Delhi, Vol4B, pp 61–76.

Ruppert, E.E. 1991. Introduction to aschelminth phyla: A consideration of mesoderm, body cavitiyes, and cuticle. In: *Microscopic Anatomy of Invertebrates*. (eds) Harrison, F.W. and Ruppert, E.E. Vol 4: Aschelminthes, Wiley-Liss, New York, pp 1–17.

Ruppert, E.E., Fox, R.S. and Barnes, R.D. 2004. Kamptozoa and Cycliophora. In: *Invertebrata Zoology*. ISBN 0-03-025982-7, pp 808–812.

Rusin, L.Yu. and Malakhov, V.V. 1998. Free-living marine nematodes possess no eutely. Dok Biol Sci, 361: 331–333.

Ruttner-Kolisko, A. 1963. The interrelationship of the Rotatoria. In: *The Lower Metazoa*. (ed) Dougherty, E.C., University of California Press, Berkeley, pp 263–272.

Ruttner-Kolisko, A. 1969. Kreuzung experimente zwischen *Brachionus urceolaris* und *Brachionus quadrideutatus*, ein Beitrage zur Fortpflanezungsbiologie der hetergonen Rotataria. Arch Hydrobiol, 65: 397–412.

Saari, S., Nareaho, A. and Nikander, S. 2019. Acanthocephala (Thorny-headed worms). In: *Canine Parasites and Prasitic Diseases*. (eds) Nareaho, A. and Saari, S., Elsevier, Amsterdam, pp 151–155.

Saito, Y. and Watanabe, H. 1982. Colony specificity in the compound ascidian, *Botryllus scalaris*. Proc Jap Acad, 58B: 105–108.

Sameoto, D.D. 1971. Life history, ecological production and an empirical mathematical model of population of *Sagitta elegans* in St. Margaret's Bay, Nova Scotia. J Fish Res Board Can, 28: 971–985.

Santagata, S. 2015a. Phoronida. In: *Evolutionary Developmental Biology of Invertebrates.* (ed) Wanninger, A., Springer Verlag, Wein. Vol 2. pp 231–246.

Santagata, S. 2015b. Ectoprocta. In: *Evolutionary Developmental Biology of Invertebrates.* (ed) Wanninger, A., Springer Verlag, Wein. Vol 2. pp 247–262.

Santagata, S. 2015c. Brachiopoda. In: *Evolutionary Developmental Biology of Invertebrates.* (ed) Wanninger, A., Springer Verlag, Wein. Vol 2, pp 263–278.

Santo, N., Caprioli, M., Rosenigo, S. and Ricci, C. 2001. Egg size and offspring fitness in a bdelloid rotifer. Hydrobiologia, 446/447: 71–77.

Sarma, S.S.S., Larios, P.S. and Nandini, S. 2001. Effect of three food types on the proportional growth of *Brachionus calyciflorusus* and *Brachionus patulus* (Rotifera: Brachionidae). Rev Biol Trop, 49n. 1 San Jose Mar. 2001.

Sarma, S.S.S., Nandini, S. and Gulati, R.D. 2002. Cost of reproduction in selected species of zooplankton (rotifers and cladocerans). Hydrobiologia, 481: 89–99.

Sasal, P., Jobet, E., Faliex, E. and Morand, S. 2000. Sexual competition in an acanthocephalan parasite of fish. Parasitology, 120: 65–69.

Sasal, P., Faliex, E., De Buron, I. and Morand, S. 2001. Sex discriminatory effect of the acanthocephalan *Acanthocephaloides propinquus* on a gobiid fish *Gobius bucchichii*. Parasite, 8: 231–236.

Sasi, H. and Giannetto, D. 2016. First record of adult Nematomorpha *Gordius* sp from Western Anatolia (Turkey). Turk J Zool, 40: 320–323.

Schiemer, F., Duncan, A. and Klekowski, R.Z. 1980. A bioenergetic study of a benthic nematode *Plectus palustris* deMan 1880 throughout its life cycle. Oecologia, 44: 205–212.

Schiemer, F. 1987. Nematoda. In: *Animal Energetics.* (eds) Pandian, T.J. and Vernberg, F.J., Academic Press, Vol 1: 185–215.

Schierwater, B. 2005. My favorite animal, *Trichoplax adhaerens.* BioEssays, 27: 1294–1302.

Schierwater, B. 2013. Placozoa, Plattentiere. In: *Spezielle Zoologie: Teil 1: Einzeller, und Wirbellose Tiere.* (eds) Westheide, W. and Rieger, R., Springer-Spektrum, Berlin, pp 103–107.

Schierwater, B. and Eitel, M. 2015. Placozoa. In: *Evolutionary Developmental Biology of Invertebrates.* (ed) Wanninger, A., Springer Verlag, Wein. Vol 1, pp 107–114.

Schiffer, P.H., Robertson, H.E. and Telford, M.J. 2018. Orthonectids are highly degenerated annelid worms. Curr Biol, 28: 1–5.

Schleip, W 1934. Die Regeneration des Russels *Phascolion strombi* Mont (Sipunculidae). Z Wiss Zool, 145: 462–496.

Schleip, W. 1935. Die Regenerationsvorgage nach Amputation des Huntereudes von *Phascolosoma minutum.* Ztschr Wiss Zool, 147: 59–76.

Schmidt, G.D. 1985. Biology of Acanthocephala. In: *Development and Life Cycles.* (eds) Nickol, B.B. and Crompton, D.W.T. Cambridge Univ Press, Cambridge, pp 273–306.

Schmidt-Rhaesa, A. 1997. *Nematomorpha.* Gustav Fischer, Cornell University, p 124.

Schmidt-Rhaesa, A. 2009. Considerations on the genus *Gordius* (Nematomorpha, horsehair worms) with the description of seven new species. Zootaxa, 2533: 1–35.

Schmidt-Rhaesa, A., Farfan, M.A. and Bernard, E.C. 2009. First record of millipeds as hosts for horsehair worms (Nematomorpha) in North America. Northeastern Nat, 16: 125–130.

Schmidt-Rhaesa, A. 2013. Nematomorpha: In: *Handbook of Zoology.* (ed) Schmidt-Rhaesa, A., De Gruyter, Berlin, Vol 1, pp 29–145.

Schmidt-Rhaesa, A. 2014. Phylum Gnathostomulida. In: *Handbook of Zoology.* (eds) Sterrer, W. and Sorensen, M.V., De Gruyter, Berlin, Vol 3: 135–196.

Schokraie, E., Hotz-Wagenblatt, A., Warnken, U. et al. 2011. Investigating heat shock proteins of tardigrades in active versus anhydrobiotic state using shotgun proteomics. J Zool Syst Evol Res, 49: 111–119.

Schroder, T., Howard, S., Arroyo, M.L. and Walsh, E.J. 2007. Sexual reproduction and diapauses of *Hexarthra* sp (Rotifera) short-lived in the Chihuahuan Desert. Freshwat Biol, 52: 1033–1042.

Schuchert, C. 1911. Paleogeographic and geologic significance of recent Brachiopoda. Geo Soc Am Bull, 22: 258–275.

Schultz, E. 1903. Aus dem Gebiete der Regeneration. IV. Uber Regenerationserscheinungen bei *Actinotrocha brachiata* Muller. Zeit Wiss Zool, 75: 473–494.

Schulze, A. 2004. Sipuncula. In: *Animal Life Encyclopedia.* (eds) Grzimek, B., Kleiman, D.G., Geist, V. and McDade, M.C., Thomson-Gale, Detroit, Vol 2, p 569.

Schulze, A., Cutler, E.B. and Giribet, G. 2004. Reconstructing the phylogeny of the Sipuncula. Hydrobiologia, 535-536: 277–296.

Schulze, F.E. 1883. *Trichoplax adhaerens,* nov. gen. nov. spec. Zool Anz, 6: 92–97.

Schwarz, E.M. 2017. Evolution: A parthenogenetic nematode shows how animals become sexless. Curr Biol, 27R: 1064–1066.

Seed, R. and Hughes, R.N. 1992. Reproductive strategies of epialgal bryozoans. Invert Reprod Dev, 22: 1–3.

Segers, S. 2008. Global diversity of rotifers (Rotifera) in freshwater. Hydrobiologia, 595: 49–59.

Seki, K. and Toyoshima, M. 1998. Preserving tardigrades under pressure. Nature, 395: 853–854.

Self, I.T. 1990. Pentastomida. In: *Reproductive Biology of Invertebrates.* (eds) Adiyodi, K.G. and Adiyodi, R.G., Oxford & IBH Publishers, New Delhi, Vol 4B, pp 157–163.

Selys-Longchamps, M. 1907. *Phoronis.* Fauna und Flora des Golfes von Neapal, 30: 1–280.

Serra, M. and Carmona, M.J. 1993. Mixis strategies and resting egg production of rotifers living in temporally-varying habitats. Hydrobiologia, 255/256: 117–126.

Serra, M., Carmona, M.J. and Miracle, M.R. 1994. Survival analysis of three clones of *Brachionus plicatilis* (Rotifera). Hydrobiologia, 277: 97–105.

Serra, M. and King, C.E. 1999. Optimal rates of bisexual reproduction in cyclical parthenogens with density-dependent growth. J Evol Biol, 12: 263–271.

Shah, M.M. and Mohilal, N. 2012. Density and diversity of nematodes infecting certain insects in Manipur. Proc Symp Biodiv Conserv Strat Nitro Professional, ISBN: 978-81-923343-1-8.

Shakya, P., Jayraw, A.K., Jamra, N. et al. 2017. Incidence of gastrointestinal nematodes in goats in and around Mhow, Madhya Pradesh. J Parasit Dis, 41: 963–967.

Shelley, A.J. 1975. A preliminary survey of the prevalence of *Mansonella ozzardi* in some rural communities on the river Purus, State of Amazonas, Brazil. Ann Trop Med Parasitol, 69: 407–412.

Shenkar, N. and Swalla, B.J. 2011. Global diversity of Ascidiacea. PLoS ONE, 6: e20657, doi:10.137/journal.pone.0020657.

Shimotori, T. and Goto, T. 2001. Developmental fates of the first four blastomeres of the chaetognath *Paraspadella gotoi*: relationship to protostomes. Dev Growth Differ, 43: 371–382.

Shostak, A.W. and Dick, T.A. 1987. Individual variability in reproductive success of *Triaenophorus crassus* Forel (Cestoda: Pseudophyllidae), with comments on use of the Lorenz curve and Gini coefficient. Can J Zool, 65: 2878–2885.

Signorovitch, A., Hur, J., Gladyshev, E. and Meselson, M. 2015. Allele sharing and evidence for sexuality in a mitochondrial clade of bdelloid rotifers. Genetics, 200: 581–890.

Silen, L. 1977. Polymorphism. In: *Biology of Bryozoans.* (eds) Wollacott, R.M. and Zimmer, R.L., Academic Press, New York, pp 184–232.

Sinisalo, T., Poulin, R., Hogmander, H. et al. 2004. The impact of sexual selection on *Corynosoma magdaleni* (Acanthocephala) infrapopulations in Saimaa ringed seals (*Phoca hispida saimensis*). Parasitology, 128: 179–185.

Skold, H.N., Ohst, M. and Skold, M. 2009. Stem cells in asexually reproducing marine invertebrates. In: *Stem Cells in Marine Organisms.* (eds) Rinkevich, B. and Matranga, V. Springer Verlag, Dordrecht, pp 105–138.

Sliusarev, G.S. 2008. Phylum Orthonectida: morphology, biology, and relationships to other multicellular animals. Zh Obshch Biol, 69: 403–427.

Snell, T.W. 1987. Sex, population dynamics and resting egg production in rotifers. Hydrobiologia, 144: 105–111.

Snell, T.W. and King, C.E. 1997. Lifespan and fecundity patterns in rotifers: the cost of reproduction. Evolution, 31: 882–890.

Snell, T.W., Kubanek, J., Carter, W. et al. 2006. A protein signal triggers sexual reproduction in *Brachionus plicatilis* (Rotifera). Mar Biol, 149: 763–773, DOI: 10.1007/s00227-006-0251-2.

Sommer, R.J. 2015. Nematoda. In: *Evolutionary Developmental of Invertebrates: Ecdysozoa I: Non-Tetraconata.* (ed) Wanninger, A., Springer Verlag, Wein. Vol 3, pp 15–34.

Sorensen, M.V., Accogli, G. and Hansen, J.G. 2010. Postembryonic development of *Antygomonas incomitata* (Kinorhyncha: Cyclorhagida). J Morphol, 271: 863–882.

Starkweather, P.L. 1987. Rotifera. In: *Animal Energetics*. (eds) Pandian, T.J. and Vernberg, F.J., Academic Press, Vol 1: 159–183.

Stelzer, C.P. 2002. Phenotypic plasticity of body size at different temperatures in a planktonic rotifer: mechanisms and adaptive significance. Funct Ecol, 16: 835–841.

Stelzer, C.-P. 2005. Evolution of rotifer life histories. Hydrobiologia, 546: 335–346.

Sterrer, W. 1974. Gnathostomulida. In: *Reproduction of Marine Invertebrates*. (eds) Giese, A.S. and Pearse, J.S., Academic Press, New York, Vol 1, pp 345–389.

Stocker, L.J. 1991. Effects of size and shape of colony on rates of fission, fusion, growth and mortality in a subtidal invertebrate. J Exp Mar Biol Ecol, 149: 161–175.

Stocker, L.J. and Underwood, A.J. 1991. The relationship between the presence of neighbours and rates of sexual and asexual reproduction in a colonial invertebrate. J Exp Mar Biol Ecol, 149: 191–205.

Storch, V. and Alberti, G. 1978. Ultrastructural observations on the gills of polychaetes. Helgoland Wiss Meeresunters, 31: 169–179.

Strayer, D.L., Hummon, W.D. and Hochberg, R. 2010. Gastrotricha. In: *Ecology and Classification of North American Freshwater Invertebrates*. (eds) Thorp, J.H. and Covich, A.P., Elsevier, Amsterdam, pp 163–172.

Streit, A. 2008. Reproduction in *Strongyloides* (Nematoda): a life between sex and parthenogenesis. Parasitology, 135: 285–294.

Stromberg, P.C. and Crites, J.L. 1974. The life cycle and development of *Camallanus oxycephalus* Ward and Magath, 1916 (Nematoda: Cammallanidae). J Parasitol, 60: 117–124.

Struck, T.H., Schult, N., Kusen, T. et al. 2007. Annelid phylogeny and the status of Sipuncula and Echiura. BMC Evol Biol, 7: 57, https://doi.org/10.1186/1471-21-148-7-57.

Stull, J.K., Haydock, C.I. and Montagne, D.E. 1986. Effects of *Listriolobus pelodes* (Echiura) on coastal shelf benthic communities and sediments modified by a major California wastewater discharge. Estu Coast Shelf Sci, 22: 1–17.

Sugiyama, N., Iseto, T., Hirose, M. and Hirose, E. 2010. Reproduction and population dynamics of the solitary entoproct *Loxosomella plakorticola* inhabiting a desmosponge, *Plakortis* sp. Mar Evol Prog Ser, 415: 73–82.

Sunnucks, P., Curach, N.C., Young, A. et al. 2000. Reproductive biology of the onychophoran *Euperipatoides rowelli*. J Zool Lond, 250: 447–460.

Svane, I. 1983. Ascidian reproductive patterns related to long-term population dynamics. Sarsia, 68: 249–255.

Szalai, A.J. and Dick, T.A. 1989. Differences in numbers and inequalities in mass and fecundity during the egg-production period for *Raphidascaris acus* (Nematoda: Anisakidae). Parasitology, 98: 489–495.

Szostakowska, B., Myjak, P., Wyszynski, M. et al. 2005. Prevalence of anisakin nematodes in fish from Southern Baltic Sea. Pol J Microbiol, 54: 41–45.

Tarjuelo, I. and Turon, X. 2004. Resource allocation in ascidians: reproductive investment vs. other life history traits. Invert Biol, 123: 168–180.

Telesh, I.V., Rahkola, M. and Vijanen, M. 1998. Carbon content of some freshwater rotifers. Hydrobiologia, 387-388: 355–360.

Teuchart, G. 1968. Zur Fortpflanzung und Entwickluing der Macrodasyoidea (Gastrotricha). Z Morphol Tiere, 63: 343–418.

Thane, A. 1974. Rotifera. In: *Reproduction of Marine Invertebrates*. (eds) Giese, A.C. and Pearse, J.S., Academic Press, New York, Vol 1, pp 471–484.

Thomas, F., Schmidt-Rhaesa, A., Martin, G. et al. 2002. Do hairworms (Nematomorpha) manipulate the water seeking behavior of their terrestrial hosts? J Evol Biol, 15: 356–361.

Thompson, A.B. 1985. Transmission dynamics of *Profilicollis botulus* (Acanthocephala) from crabs (*Carcinus maenas*) to eider ducks (*Somateria mollissima*) on the Ythan Estuary, N.E. Scotland. J Anim Ecol, 54: 605–616.

Thorp, D. and Roger, D.C. 2014. *Ecology and General Biology: Thorp and Covich's Freshwater Invertebrates*. Academic Press, London, p 1148.

Todaro, M.A. and Shirley, T.C. 2003. A new meiobenthic priapulid (Priapulida, Tubiluchidae) from a Mediterranean submarine cave. Ital J Zool, 70: 79–87.

Tyler, A. and Bauer, H. 1937. Polar body extrusion and cleavage in artificially activated eggs of *Urechis caupo*. Biol Bull, 73: 164–180.

Uchida, T. and Yamada, M. 1983. Cnidaria. In: *Invertebrate Embryology*. (eds) Dan, K., Sekiguchi, K., Ando, Y. and Watanabe, H., Baifukan, Tokyo, Vol 1, pp 103–133.

Umbers, K.D.L., Byatt, L.J., Hill, N.J. et al. 2015. Prevalence and molecular identification of nematode and dipteran parasites in an Australian Alpine grasshopper (*Kosciuscola tristis*). PLoS ONE, 10: e0121685.

Vincent, A.G. and Font, W.F. 2003. Host specificity and population structure of two exotic helminthes, *Camallanus cotti* (Nematoda) and *Bothriocephalus acheilognathi* (Cestoda), parasitizing exotic fishes in Waianu stream, O'ahu, Hawai'I. J Parasitol, 89: 540–544.

Viney, M.E., Steer, M.D. and Wilkes, C.P. 2006. The reversibility of constraints on size and fecundity in the parasitic nematode *Strongyloides ratti*. Parasitology, 133: 477–483.

von Dohren, J. 2015. Nemertea. In: *Evolutionary Developmental of Invertebrates*. (ed) Wanninger, A., Springer Verlag, Wein. Vol 2, pp 155–192.

Wallace, R.L., Cipro, J.J. and Grubbs, R.W. 1998. Relative investment in offspring by sessile rotifera. Hydrobiologia, 387/388: 311–316.

Wallace, R.L. and Snell, T.W. 2010. Rotifera. In: *Ecology and Classification of North American Freshwater Invertebrates*. (eds) Thorp, J.H. and Covich, A.P., Elsevier, pp 173–235.

Walls, J.G. 1982. *Encyclopedia of Marine Invertebrates*. (ed) Neptune, N.J., T.F.H. Publications, p 736.

Walz, N., Sarma, S.S.S. and Benker, U. 1995. Egg size in relation to body size in rotifers: An indication of reproductive strategy? Hydrobiologia, 313-314: 165–170.

Wanninger, A. 2015a. *Evolutionary Developmental Biology of Invertebrates: Non-Bilateria*. Springer Verlag, Wein. Vol 1, p 258.

Wanninger, A. 2015b. *Evolutionary Developmental Biology of Invertebrates: Lophotrochozoa* (Spiralia). Springer Verlag, Wein. Vol 2, p 289.

Wanninger, A. 2015c. *Evolutionary Developmental Biology of Invertebrates: Deuterostomia*. Springer Verlag, Wein. Vol 3, p 221.

Wanninger, A. 2015d. Entoprocta. In: *Evolutionary Developmental Biology of Invertebrates*. Springer Verlag, Wein. Vol 2. pp 89–102.

Wanninger, A. and Neves, R. 2015. Cycliophora. In: *Evolutionary Developmental of Invertebrates*. Springer Verlag, Wein. Vol 2, pp 79–88.

Watanabe, M. 2006. Anhydrobiosis in invertebrates. Appl Entomol Zool, 41: 15–31.

Wegener, F. 1938. Beitrag zur Kenntnis der Russelregeneration der Sipunculiden. Z wiss Zool, 150: 527–565.

Weiss, M.J. and Levy, D.P. 1979. Sperm in "parthenogenetic" freshwater gastrotrichs. Science, 205: 302–303.

Welch, D.B.M., Welch, J.L.M. and Meselson, M. 2008. Evidence for degenerate tetraploidy in bdelloid rotifers. Proc Nat Acad Sci USA, 105: 5145–5149.

Welch, P.S. and Wehrle, L.P. 1918. Observations on reproduction in certain parthenogenetic and bisexual nematodes reared in artificial media. Trans Am Microsc Soc, 37: 147–176.

Wennberg, S. 2008. Aspects of priapulid development. Uppsala University, Dissertation, Faculty of Science and Technology, p 451.

Westh, P., Kristiansen, J. and Hvidt, A. 1991. Ice-nucleating activity in the freeze-tolerant tardigrade *Adorybiotus coronifer*. Comp Biochem Physiol A, 99: 401–404.

Whitfield, P.J. and Evans, N.A. 1983. Parthenogenesis and asexual multiplication among parasitic platyhelminths. Symp Brit Soc Parasit, 20: 121–160.

Wilczynski, J. 1968. On the sex in *Bonellia viridis*. Acta Biother, 18: 338–360.

Winston, J.E. and Jackson, J.B.C. 1984. Ecology of cryptic coral reef communities. IV. Community development and life histories of encrusting cheilostome Bryozoa. J Exp Mar Biol Ecol, 76: 1–21.

Woo, P.T.K. 2006. Fish Diseases and Disorders: Protozoan and Metazoan Infections. GABI International, UK, Vol 1, p 802.

Wood, T.S. 2007. Development and metamorphosis of cyphonautes larvae in the freshwater ctenostome bryozoans, *Hislopia malayensis* Annandale, 1916. Proc 14th Int Bryozool Asso Conf, Virgenia. Mus Nat His Spec Publ, 15: 339–346.

Wood, T.S. 2010. Bryozoans. In: *Ecology and Classification of North American Freshwater Invertebrates.* (eds) Thorp, J.H. and Covich, A.P., Elsevier, Amsterdam, pp 437–454.

Woombs, M. and Laybourn-Parry, J. 1984. Growth, reproduction and longevity in nematodes from sewage treatment plants. Oecologica, 64: 168–172.

Wurdak, E.S. and Gilbert, J.J. 1976. Polymorphism in the rotifer *Asplanchna sieboldi*: Fine structure of saccate, cruciform and companulate females. Cell Tiss Res, 169: 435–448.

Yadav, A.K., Schmidt-Rhaesa, A., Laha, R. and Sen, A. 2017. On the recovery of horsehair worms, *Gordius* sp (Nematomorpha: Gordiida) from pork in Shillong, India. J Parasit Dis, 41: 302–304.

Yadav, B.C., Veluthambi, K. and Subramaniam, K. 2006. Host-generated double stranded RNA induces RNAi in plant-parasitic nematodes and protects the host from infection. Mol Biochem Parasitol, 148: 219–222.

Yoshinaga, T., Hagiwara, A. and Tsukamoto, K. 2003. Life history response and age-specific tolerance to starvation in *Brachionus plicatilis* O.F. Muller (Rotifera). J Exp Mar Biol Ecol, 287: 261–271.

Yusa, Y. 1996. Utilization and degree of depletion of exogenous sperm in three hermaphroditic sea hares of the genus *Aplysia* (Gastropoda: Opisthobranchia). J Mol Stud, 62: 113–120.

Zaika, V.E. and Makarova, N.P. 1979. Specific production of free-living marine nematodes. Mar Ecol Prog Ser, 1: 153–158.

Zervos, S. 1988. Population dynamics of a thelastomatid nematode of cockroaches. Parasitology, 96: 353–368.

Zhang, Z-Q. 2011. Animal biodiversity: An introduction to higher-level classification and taxonomic richness. Zootaxa, 3148: 7–12.

Zhao, L., Zhang, S., Wei, W. et al. 2013. Chemical signals synchronize the life cycles of a plant-parasitic nematode and its vector beetle. Curr Biol, 23: 2038–2043.

Zimmer, R.L. 1964. The morphology and function of accessory reproductive glands in the lophophore *Phoronis vancouverensis* and *Phoropsis harmeri*. J Morphol, 121: 159–178.

Zirpola, G. 1924. Sulla genesi della colonie primaverilli del *Zoobotryon*. Boll Soc nat Napoli, 4: 113–128.

Zurbuchen, K. 1937. Entwicklungsmechanische Untersuchungen an *Bonellia viridis*. II. Entwicklung der Intersexe und sexuelle Variabilität bei *Bonellia viridis* in Versuchen mit abgekürzten Russelparasitismus. Pubbl Staz Zool Napoli, 16: 28–79.

Zwart, R.S., Thudi, M., Channale, S. et al. 2019. Resistance to plant-parasitic nematodes in chickpea: current status and future perspectives. Front Plant Sci, 10: 966, DOI: https://doi.org/10.3389/fpls.2019.00966.

Author Index

Species Index

Subject Index

Author's Biography

Recipient of the S.S. Bhatnagar Prize, the highest Indian award for scientists, one of the ten National Professorships, T.J. Pandian has served as editor/ member of editorial boards of many international journals. His books on Animal Energetics (Academic Press) identify him as a prolific but precise writer. His five volumes on Sexuality, Sex Determination and Differentiation in Fishes, published by CRC Press, are ranked with five stars. He is presently authoring a multi-volume series on Reproduction and Development in Aquatic Invertebrates, of which the volumes on Crustacea, Mollusca, Echinodermata and Prochordata, Annelida, and Platyhelminthes have already been published. The next one, Minor Phyla, is in your hands.

Author's Biography

Recipient of the S.S. Bhatnagar Prize, the highest Indian award for scientists, one of the ten National Professorships, T.J. Pandian has served as editor/ member of editorial boards of many international journals. His books on Animal Energetics (Academic Press) identify him as a prolific but precise writer. His five volumes on Sexuality, Sex Determination and Differentiation in Fishes, published by CRC Press, are ranked with five stars. He is presently authoring a multi-volume series on Reproduction and Development in Aquatic Invertebrates, of which the volumes on Crustacea, Mollusca, Echinodermata and Prochordata, Annelida, and Platyhelminthes have already been published. The next one, Minor Phyla, is in your hands.